SUSTAINABILITY

WHAT EVERYONE NEEDS TO KNOW®

SUSTAINABILITY

WHAT EVERYONE NEEDS TO KNOW®

PAUL B. THOMPSON AND
PATRICIA E. NORRIS

OXFORD
UNIVERSITY PRESS

OXFORD
UNIVERSITY PRESS

Oxford University Press is a department of the University of Oxford. It furthers
the University's objective of excellence in research, scholarship, and education
by publishing worldwide. Oxford is a registered trade mark of Oxford University
Press in the UK and certain other countries.

"What Everyone Needs to Know" is a registered trade mark of
Oxford University Press.

Published in the United States of America by Oxford University Press
198 Madison Avenue, New York, NY 10016, United States of America.

© Oxford University Press 2021

Library of Congress Cataloging-in-Publication Data
Names: Thompson, Paul B., author. | Norris, Patricia E., author.
Title: Sustainability : what everyone needs to know / Paul B. Thompson and
Patricia E. Norris, Michigan State University.
Description: New York, NY : Oxford University Press, [2021] |
Series: What everyone needs to know | Includes bibliographical references
and index.
Identifiers: LCCN 2020024401 (print) | LCCN 2020024402 (ebook) |
ISBN 9780190883249 (hardback) | ISBN 9780190883232 (paperback) |
ISBN 9780190883263 (epub) | ISBN 9780197525883
Subjects: LCSH: Sustainable development. | Sustainability.
Classification: LCC HC79.E5 T484 2021 (print) | LCC HC79.E5 (ebook) |
DDC 304.2—dc23
LC record available at https://lccn.loc.gov/2020024401
LC ebook record available at https://lccn.loc.gov/2020024402

1 3 5 7 9 8 6 4 2

Paperback printed by LSC Communications, United States of America
Hardback printed by Bridgeport National Bindery, Inc., United States of America

To Camille and Everett
My grandchildren (so far) who inherit these places
P. B. T.

To George and Ellen Norris
Who taught me about caring for special people and places
P. E. N.

CONTENTS

9 Sustainability: What Everyone Needs to Ask   215

ACKNOWLEDGMENTS

This book is the product of an extended collaboration with many people who have influenced us, joined with us in allied projects, and helped us bring the manuscript to its final form. It would never have been possible but for the creation of the Sustainable Michigan Endowed Project, affectionately known locally by the acronym SMEP. SMEP was the brainchild of Sandra Batie, Emeritus Elton R. Smith Chair in Food and Agricultural Policy at Michigan State University (MSU) and the first person we need to thank. It began in 2002 when funds from Michigan State University's College of Agriculture and Natural Resources (CANR) were pledged to match a gift from the W. K. Kellogg Foundation. The project was intended to support research and graduate training related to sustainability in the context of Michigan. Sustainability was not popular then, and very few CANR researchers were willing to organize or advertise their activity under the banner of sustainability. Our anecdote about sustainable agriculture in chapter 1 dates back to those days.

At first, the main activity of SMEP was an informal seminar/workshop series that Batie organized with MSU faculty in CANR holding endowed chairs. The function of this group was to plan and guide the expenditure of SMEP funds, which required extensive discussion and brainstorming about the shape that a sustainability-oriented program might take at MSU. At the outset, in addition to Batie and Thompson, this group included Richard Bawden, Visiting Distinguished Professor; David Beede, the Clinton E. Meadows Professor in Dairy Management; Mike Hamm, the C. S. Mott Professor of Sustainable Agriculture; Chris Peterson, Homer Nowlin Chair of Consumer Responsive Agriculture; and Joan Rose, the Homer Nowlin Chair in Water Research. Within a few years, others joined the group, including Norris; Jim Detjen, Knight Professor in Environmental Journalism; Jianguo "Jack" Liu, the Rachel Carson Chair in Sustainability; Tom Dietz, Director of the Environmental Science and Policy Program; and Mark Skidmore, the Morris Chair in State and Local Government Finance and Policy. Those were the originals, but additional MSU faculty joined as some retired or new endowed positions were created. They include Rick Foster, Eric Freedman, David Hennessy, Bill Porter, Jim Tiedje, Kyle Whyte, Felicia Wu, and Jinhua Zhao. We should also note MSU colleagues who were important collaborators and influences during these early years, even if they did not participate in SMEP. They include Larry Busch, George Byrd, David Schweikhardt, and Stuart Gage. All of these people have shaped the understanding of sustainability that informs this book. This is not to say that all of them would agree with us.

The second main activity that preceded our collaboration on this book was the development of a new undergraduate course for MSU's Department of Community Sustainability. The department was created through the merger of three separate programs in 2003. Representing a wildly diverse set of academic interests and expertise, faculty worked diligently to discover where the new academic unit might locate its intellectual core. By 2010, faculty recognized that the idea of sustainability provided the basis for a common intellectual core, and the department decided to reorganize the undergraduate program around three core courses and a series of learning outcomes deemed necessary for any student of sustainability. Theoretical Foundations of Sustainability was to be the core course focusing on the principles underlying sustainability as the concept would be developed and applied in subsequent courses. We worked on the course for three years, including one year during which we taught it together. The process refined the thinking we had done in SMEP, and it also involved substantial input from colleagues in the Department of Community Sustainability: Robby Richardson, Mike Hamm, and Laura Schmitt-Olabisi, among others. Our Department of Community Sustainability has been fortunate to recruit a number of young faculty to join our grand adventure, and they have contributed mightily to advancing departmental work on sustainability and to our own understanding of this field of study: Maria Claudia Lopez, Jenny Hodbod, Lissy Goralnik, and Steven Gray, among others.

The other thing that teaching a course forced us to do was find literature that could support our teaching. Hands down, the most influential thing was Donella Meadow's posthumously published *Thinking in Systems*. Our course evolved as one that would augment Meadow's chapters on stock and flow systems modeling with enough background in ecology, economics, development theory, and applied environmental policy studies that students could see how a common notion of sustainability was running through each of these domains. Our course emphasized the systems thinking process and used Meadow's graphic representation of stocks, flows, and feedbacks in lieu of mathematical models. This reflected Thompson's training in philosophy and our agreement that moving too quickly to the quantitative representation alienates many students and prevents them from grasping what the math is actually about.

There are also people who influenced our thinking either through personal interactions or through our reading. Glenn Johnson, Dave Ervin, Bryan Norton, and Michael O'Rourke would be high on the first list, while John Gowdy, Herman Daly, Elinor Ostrom, Kenneth Boulding, C. West Churchman, Aldo Leopold, John Dewey, Jane Addams, Heȟáka Sápa, and

Henry David Thoreau are high on the second. Both lists could go on and on; we've been at this business for a while.

We began to think about a book collaboration in 2017 at the suggestion of Lucy Randall at Oxford University Press (OUP). Lucy sent us a few titles from Oxford's What Everyone Needs to Know® series and invited us to prepare a prospectus that would be consistent with the series' question-and-answer approach. We had already reviewed other books intended to provide a non-textbook introductory treatment, as well as new literature from authors suggesting that we should forget about sustainability and start working on resilience. We incorporated our thoughts on these titles into a prospectus for a book that would take the same philosophical approach as the course we had developed, while avoiding the typical vices of academic publishing. Lucy has been especially helpful as our project has moved from a proposal to a manuscript, reviewing early chapter drafts for tone and accessibility and then, with OUP's Hannah Doyle, editing the first draft of the book. We want to thank Lucy and Hannah, as well as the external reviewers of the prospectus recruited by OUP.

We also recruited a number of colleagues to read individual chapters in their areas of specialization. We would like to thank Dan McCole, Robby Richardson, Laura Schmitt-Olabisi, Eric Scorsone, and Laurie Thorp from MSU and Kurt Stephenson from Virginia Tech for valuable suggestions that have almost certainly prevented us from looking like total idiots. In the same vein, we thank Dane Scott, OUP's external reviewer, who provided comments on the entire manuscript. Of course, we take full ownership of anything we got wrong or places where readers may strongly disagree with us.

There are also people who have helped us move the book into production. Julie Eckinger provided technical assistance with file management and preparation.

We also acknowledge the financial support we have each received from MSU AgBioResearch, USDA Hatch Projects MICL02324 (Thompson) and MICL02158 (Norris). The W. K. Kellogg Foundation provided an endowment for the MSU Chair in Agricultural, Food, and Community Ethics, and the Gordon and Norma Guyer and Gary L. Seevers endowment supports the MSU Chair in Natural Resource Conservation.

SUSTAINABILITY

WHAT EVERYONE NEEDS TO KNOW®

1

WHAT IS SUSTAINABILITY?

What is sustainability?

Sustainability is a measure of whether (or to what extent) a process or practice can continue. This is a very general characterization. People flesh it out in many ways. The process or practice can be very ordinary, such as going to the grocery store to do your weekly shopping. Or it can be exceedingly complex and comprehensive: the entire system of production and exchange that makes up the global economy, for example. Sometimes you can measure how long or to what degree something can continue by collecting data. Other times, you just estimate whether something is sustainable, making a seat-of-the-pants kind of judgment. One way or another—whether precisely or generally—people can analyze sustainability for an exceedingly large number of different activities. We can assess the sustainability of a particular farming method or of the entire food sector. We can determine the sustainability of one particular building or architectural design, but we can also look at the built environment for a city. Given the right data and theoretical tools, one can evaluate the sustainability of a company's business practice or an entire sector of the economy. Specialists can compute ratings for the sustainability of alternative types of packaging and even apply the concept to natural processes, such as a population of organisms or a volcanic eruption.

While people talk meaningfully about the sustainability of many different processes or practices, in this book we emphasize sustainability of systems. Big systems, like the economy or a regional ecosystem, are composed of smaller-scale practices that affect one another. Trying to understand how they fit together leads one to think about connections and connectedness. Or to put it another way, it leads us to understand seemingly isolated and unconnected activities or processes as occurring within a system. The answer to questions about whether any process or practice can continue depends on the larger system in which it is embedded. Continuity of the smaller systems it depends on also matters. This has led people who focus on the sustainability of a business enterprise, a community, an ecosystem, or a way of life to develop a general approach to systems thinking: conceptualizing things in terms of the larger and smaller systems on which they rely and probing the ways in which seemingly disparate activities and happenings are connected to one another.

The flexibility with which systems can be described leaves some people dazed and confused. One person may have a very narrow understanding of a given practice or process, while someone else may understand it more broadly. If one car gets better gas mileage than another one, for example, it can be said to be more sustainable because it goes further on a gallon of fuel. But some would say that *any* combustion of fossil fuel is unsustainable because our supply of petroleum is finite and will eventually run out. Even more generally, the idea of sustainability has been advanced as a broad, comprehensive social goal over the last four decades. In this sense, people are thinking about a large interconnected set of practices that are fundamental to our current way of life, and sustainability is a measure of whether (or to what extent) that way of life can continue. If this larger social context is what someone has in mind, the sustainability of a more specific practice or process (such as driving a car with low or high gas mileage) has to be assessed in terms of whether it promotes the continuance of a particular way of life or inhibits it.

Getting a handle on what makes a particular practice sustainable is useful for people who have an interest in it. Yet sustainability would not have become a buzzword except for the way that it often implies something about the totality of practices and processes on which everyone depends. This is the sense in which sustainability is a *big idea*. To get a handle on sustainability as a big idea, one must imagine all the different meanings and activities associated with the way people live today and then ask how they connect to form a total system of practices and processes. Like most big ideas, what any given individual thinks is important reflects that person's experiences and life goals. When people start to think about things that connect to our way of life, they start from different places. Differences among individuals' starting points—their experiences and life goals—explain much of the disagreement about what sustainability is.

If everything is interconnected with everything else, where do you start?

This is a challenge. You could say that it doesn't matter much where you start in systems thinking, because thinking through connections will eventually lead you to consider the larger context. But choosing a starting point is important for launching a conversation (or writing a book) because the entry point into sustainability that is obvious for one person will seem obscure (or boring) to someone else. We (or most of us) care deeply about the continuation of *some* things that can be understood in systems terms: the company we work for, the community we live in, the church we attend, the forests and streams where we like to hike, camp, or fish. But people may not immediately consider how all these things are connected to each other or how continuing one of them might affect another. Some people *do* think comprehensively about the earth as one big system, but the planet earth may continue even as places and institutions within it that other people care about vanish. However, you can learn a lot about the principles of sustaining a church, a community, or a favorite natural area by examining what it takes to sustain a business—or at least that's the premise of our book. We start by thinking about what it takes to sustain a business. Getting the hang of these principles is what we, your authors, think everyone should know about sustainability.

Is sustainability always about the environment?

As will become clear when we start to look at some examples, many processes and practices draw upon natural resources or depend upon services produced by the earth's ecosystems. It would be meaningless to estimate the sustainability of such practices without taking their use of and effects on natural systems into account. What is more, much of the impetus for interest in sustainability derives from growing awareness of natural resource depletion and damage to the quality of the water people drink and the air that people breathe. Sustainability is, in many people's minds, primarily defined in terms of environmental impact.

Yet there are other ways in which people evaluate sustainability that are only indirectly related to the environment. As will become clear later, part of the reason sustainability became a watchword in international politics was that poorer countries were resisting global initiatives to constrain economic development in the interest of environmental protection. They were more comfortable with the idea of *sustainable development* precisely because they saw this notion as recognizing the priority of meeting crucial human needs. Even more generally, the sustainability of some practices depends on whether they have a secure financial base. When administrators in government, universities, charitable foundations, and other large organizations ask whether a new program will be sustainable, they are often interested in whether it will generate an on-going source of its own financial support (through user fees or client payments, for example). If they will have to keep putting more and more money into it in order to keep it going, they will judge the program unsustainable. So, despite widespread suppositions to the contrary, there are many ways in which sustainability is *not* just about the environment. It is a concept applied to evaluate the regenerative capacity of many different processes, including some that have little relationship to natural resources or ecosystem services.

Is sustainability primarily about climate change?

Many people who started with a concern for environmental quality are now shifting their focus to the way that greenhouse gas emissions threaten permanent changes to earth's climate. There is little doubt that changes to average temperature and concomitant effects of melting in the polar icecaps are already affecting the stability of many global ecosystems and threatening the lives of wild species. Farmers are struggling to cope with these changes, and if predicted shifts in rainfall, drought, and temperature materialize, the planet's ability to provide enough food for its human population will be challenged in a matter of a few decades. Effects of this kind certainly sound like problems for sustainability. Some authors have placed this concern right at the forefront of sustainability thinking.

While we don't doubt the significance of climate change for sustainability, climate change isn't our focus here. If you wish to read about that important topic in the same question-and-answer format we use here, we recommend *Climate Change: What Everyone Needs to Know*® by Joseph Romm, published in the same series as this book. We focus on explaining how the idea of sustainability draws upon more familiar kinds of interconnected processes that can be found in many domains, not just the atmospheric flows that are driving climate change. We hope the answers that we give to questions about sustainability will equip readers to make connections between the impacts of climate change and dimensions of sustainability that derive from business practice, the challenges of governance, economic development, and other social domains where concepts of sustainability are applicable.

Does progress in the economy, society, and the environment add up to sustainability?

A simple diagram with three overlapping circles is often used to represent sustainability. The circles are sometimes labeled *society*, *environment*, and *economy* or *people*, *planet*, and *profit*. We call this *three circle sustainability*. Sometimes the three domains are described as the three pillars of sustainability, but the idea is the same. As we show in chapter 2, this model may derive from triple bottom line discussions in business. The good thing about this model is that it takes people beyond the environment, suggesting that other areas matter too. But it has weaknesses. First, it can easily be understood to imply simply that efforts to protect the environment have to allow businesses to make profits and people to meet needs. Profitability and social progress are viewed as constraints on environmental sustainability, but not necessarily as activities that can be more or less sustainable in themselves. Second, although it suggests that all three domains must be satisfied, it does not tell us how or even whether they are interconnected. Are there, for example, business practices that feed back to the environment, and do things happening in the environment impact the sustainability of business? The three-circle diagram does not encourage us to think that such connections are crucial for considering sustainability. Third, the tendency to focus on the nexus of the three circles as some sort of sweet spot overlooks the likelihood that sustainability decisions may require significant trade-offs among goals in each domain. Finally, the notion of three domains may wind up being limited. Are taxation and government finance in the society circle or the economy circle?

Or have these practices been left out altogether? What about scientific research, religion, and the arts? Are they adequately accounted for by the term *society*? Although we discuss activities that could be placed within one of these three domains, we prefer to think of sustainability as a way to imagine and evaluate a much broader array of practices and processes.

Is sustainability always a good thing?

The question of whether a process or practice can continue is understood in multiple ways. Someone might think of this as a simple matter of fact: either the practice will in fact keep going, or some limiting factors will cause it to decline, degrade, or come to a stop. Yet sometimes the word *can* implies permission, and saying that one cannot continue to do something means that they will not be allowed to do it. Of course, these two meanings of *can* sometimes conflict with one another. The teacher at the front of the room bangs the desk and says, "This behavior simply cannot continue," while the smart alecks in the back row are saying to one another, "Why not? We can keep it up indefinitely!" With respect to sustainability, this means there is a built-in tension. Conceptualizations that stress the underlying feasibility of the process or practice coexist with those that emphasize its desirability or permissibility. Sometimes, in other words, sustainability reflects a judgment about whether a process or practice *should* continue.

Some highly undesirable phenomena certainly seem to be all too sustainable. War, poverty, disease, and human misery show no signs of ceasing any time soon. Although it is logically possible to ask whether some evil is sustainable in the factual sense of being continued into the future, this is not how the concept of sustainability is used by people who are stressing the goodness or permissibility of continuing. Indeed, when talking about the comprehensive system of practices that supports our general way of life, people typically assume that sustainability implies progress in dampening or eradicating vices such as prejudice, oppression, or injustice. Here again there are opportunities for disagreement and misunderstanding, because people differ in the way they perceive progressive social goals. This may be a different kind of disagreement than the one arising from different starting points. Some disagreements over the meaning of sustainability have their root in disagreements about what is possible and desirable in social life.

This tension between measuring whether a process can continue and recommending that it should (or should not) continues to pop up in several chapters. We address it directly in chapter 6 in the context of social justice. More generally, we do not recommend tying your idea of sustainability too closely to specific social or ethical goals. On the one hand, the act of promoting sustainability emphasizes the good things people want to sustain. On the other, understanding how evils persist is a prerequisite for change. Someone can use the idea of sustainability to support both types of thinking. In our approach, sustainability is an umbrella term that covers a diverse set of social and natural processes, not all of which are unambiguously good. So no, sustainability is not always a good thing.

Is sustainability a social movement?

A social movement is a large and wide-ranging group action generally intended to bring about significant change in social norms and expectations. The means by which change might be pursued are variable; sometimes they require formal actions by political parties or organizations (such as labor unions), while other times they are informal and occur through changes of attitude. Social movements may—and indeed generally do—encompass a number of mutually incompatible objectives. They are often easier to identify after they are over rather than while they are occurring. Some social movements succeed in achieving key objectives, and others fail. Besides the labor movement, other key twentieth-century social movements include the civil rights movement and the women's movement.

Undoubtedly, some people view sustainability as a program of very comprehensive and even radical social change. However, given our starting characterization of sustainability as a measure of the extent to which any process or practice might continue, it can be applied in many situations that have very little to do with social change. During the first two-thirds of this book, we emphasize concepts and methods that have very wide application and can be (and indeed are being) deployed by actors and organizations that have little interest in institutional change. These include businesses, some of which are very large multinational corporations with a significant interest in maintaining the status quo. In short, we think that viewing sustainability as a social movement suffers from the same problem as assuming that sustainability is always a good thing. It gets in the way of thinking about underlying or more comprehensive systems that support both good things and bad things.

Is sustainability opposed to economic growth?

No. In fact, much of the technical research on sustainability has been developed within the economic theory of growing economies. As we discuss in chapter 5, growth is a key criterion for economic development, but it is not the only criterion. For the time being, it is important simply to recognize that economic growth can take many forms. While some forms of growth can cause the depletion of resources and irreversible damage to earth's ecosystems, other forms of growth use very few resources and actually restore ecosystems. Sometimes growth helps everyone, but other types of growth encourage social instability, warfare, and revolution. The question, then, is what kind of growth we are talking about. A more detailed answer to that question requires some setup discussion on the measurement of growth and its relationship to development. All of that is the subject of chapter 5.

Does sustainability imply a political agenda?

Again, no, though many people who promote sustainability do have political agendas. Very broad definitions of sustainability can be filled in with differing ideological viewpoints, cultural assumptions, and political opinions. Nevertheless, the idea of sustainability is not in itself inherently tied to one political perspective or another. Nor is it subjective or simply a matter of opinion. We recall a friend who, in a discussion with members of a farm organization, was asked whether he supported sustainable agriculture. It was a loaded question reflecting the view that sustainable agriculture is a political attack on mainstream farm practices. Given the context, being critical of mainstream agriculture was not politically acceptable. Our friend responded that he certainly didn't support unsustainable agriculture. The questioner was associating sustainability with a particular political orientation, but he was called up short by the suggestion that he might be advocating unsustainable agriculture. The underlying issue is whether any form of agriculture itself is sustainable rather than the type of food system one supports for political reasons. The lesson here is that thinking about the sustainability of a practice or process has a factual basis, even when there may be sharp political divisions about the practice or process itself.

When we get down to specific cases, it is often possible to identify the underlying structure and requirements of a process or practice and develop genuine insight into what allows it to continue or be reproduced from one period of time to the next. There is no recipe for sustainability that is applicable to every case, but as we move from context to context in this book, we consider ways to ascertain the sustainability of certain practices in a manner that is objective and factually based.

Our friend's answer to the sustainable agriculture question also shows that beliefs about *which* processes or practices should be sustained may indeed depend upon a person's political orientation. The approach we take in this book is to illustrate some of the ways that methods and criteria have been developed for determining the sustainability of business practices, public policies, and ways in which human beings affect the broader natural environment. In considering these examples, we usually focus on the more factual sense of "can continue." Conceptualizations of sustainability that stress a normative evaluation of whether something should continue are especially inclined to be described as articulating a social goal. In situations where sustainability is being put forward as an agenda for changing social practices in order to make them more fair, just, or equitable, we explicitly indicate that by using phrases such as *social goal* or *justice*.

Is sustainability achievable?

Nothing lasts forever, as the saying goes. There are waggish types who move from that thought to the conclusion that sustainability is a logical impossibility. Yet who said that a practice or process has to continue throughout eternity in order to be considered sustainable? There are many ways to measure how long or to what extent a practice can endure, given its background conditions. There is no reason to think that it is not sustainable just because there is some dramatic change in those conditions. This is the "What would happen if the earth was hit by a comet?" thought experiment. Current thinking is that a mass extinction event occurred when some large celestial object struck our planet about sixty-five million years ago. The climatic disruption led to the extinction of an estimated 50 percent of the species in existence at that time. It is not clear what someone might mean by saying that these extinct species had an unsustainable life process because they did not survive a meteor impact.

Some environmental scientists are predicting that people alive today could see a comparable loss of species diversity within their lifetime. As noted already, some scientific models of climate change include scenarios that would make the planet uninhabitable by many current life forms (including human beings). Such a dramatic change in the background conditions for life on earth really would render many of the ongoing attempts to be more sustainable meaningless. Religious and philosophical traditions speculate about the possibility of an end time, an Armageddon, an apocalypse, or a return to chaos. In some traditions, these events are viewed positively, while in others they are stages in a recurring cosmic cycle. They always imply that much of what makes our lives meaningful in the present is subject to destruction, upheaval, and desolation. Both scientific and theological-cosmological questions about whether sustainability is achievable thus deserve a response. So perhaps the question is not so silly after all.

We don't presume to address whether sustainability is achievable in such cosmic or ultimate terms. It is still meaningful to examine each step on life's journey and ask whether someone might be damaging the systems on which everyone depends. This points us away from thinking about sustainability as an end point, as something that could be accomplished once and for all. Indeed, what specialists learn from looking at natural systems (in chapter 3) is that they respond to and change with shocks caused by things like earthquakes, hurricanes, or floods. As others adapt this lesson to human systems such as the economy, a household, or a government, one can assess whether these systems seem able to continue and respond to disruption or whether a major shock would threaten their very survival.

Scientific approaches aim to strengthen the integrity of key systems, even as they investigate the potential for their disruption. Theological and philosophical traditions claim that everyone must carry on with our daily lives, that no one should abandon the belief that their life is meaningful. Neither scientific predictions of disruption nor religious speculations on humanity's ultimate fate justify immoral behavior in the here and now. No religious or philosophical tradition has ever held that the eventual end of life as humans know it absolves people of their duty to do the best they can while the world endures. Continuing our way of life while trying to end the evils that beset us is a responsibility for the present, even *if*, in the end, everything must go. In fact, the thought of apocalyptic collapse reminds us that failures and losses are already occurring. Acknowledging this can help us appreciate that some people experienced the end of their world decades or a century ago. Industrialization and colonial expansion have so thoroughly altered the environments in which many indigenous peoples lived that they think of themselves as adapting to a post-apocalyptic situation. This does not make survival and sustainability any less of an imperative for them, but it should inject a note of humility into those who think of sustainability strictly in terms of saving the world or warding off collapse.

One should not think of sustainability as an endpoint, especially when thinking of sustainability in the "big idea" sense. Some people like to make this point by saying that sustainability is about the journey, not the destination. It is possible for the systems on which everyone depends to be more sustainable than they currently are, and it is possible to evaluate our choices in terms of their impacts on the sustainability of other practices and processes. This neither implies nor depends upon some perfect world where sustainability is a finished project.

Where did the idea of sustainability come from?

There is no single answer to this question. Environmental historians often point to the emergence of methods for determining how much timber could be harvested from a stand of trees on a continuous basis. Yet political theorists were formulating questions about the sustainability of a state or political regime for many years before these methods first appeared in German forestry. There is reason to think that the idea of sustainability has been around for a long time. Yet we can assign a date for the beginning of its recent growth in popularity. That would be 1987, the year that the World Commission on Environment and Development published its report *Our Common Future*. In subsequent discussions, this group came to be known as the Brundtland Commission, and *Our Common Future* is often called the Brundtland report. Gro Haarlem Brundtland, the former president of Norway, chaired the World Commission on Environment and Development. She is a leading figure in promoting global initiatives to address environmental issues.

The Brundtland report documented the challenge of continuing to promote economic development, especially in countries that were not enjoying the benefits of industrialization, while also maintaining environmental quality. The eventual exhaustion of finite resources (such as fossil fuels) was one of the key challenges noted in the report; the potential for damage to ecosystems that support renewable resources (such as food or clean water) was another. The Brundtland report stressed that all countries must face these challenges and achieve sustainable development. The Brundtland Commission then gave us a memorable phrase: sustainable development "meets the needs of the present without compromising the ability of future generations to meet their own needs."

The Brundtland report affected global policy and planning in a number of dramatic ways. It basically stated that when developed economies (such as the United States and Europe) deplete resources and damage ecosystems, they are failing to meet a responsibility to future generations. It also said that societies with high rates of poverty and low standards of living must be allowed to continue economic growth. The imbalance in global standards of living should not be allowed to persist. This vision stimulated a surge of research and reports on how and whether economic development activities in Africa, Asia, and Latin America could continue without violating the maxim that future generations should be able to meet their needs. This upsurge of discussion on sustainable development eventually spread into other areas of planning and policy. The basic problem identified by the Brundtland Commission was generalized to apply broadly in local and national planning, in building construction, and in activities ranging from architecture to agriculture. For many, the idea of meeting the needs of the present without compromising future generations' ability to meet their own needs came to be equated with sustainability itself.

Has the idea of sustainability changed over time?

This is not a question that can be answered in a sentence or two. The examples we discuss in the following chapters illustrate that shifting from one practice or process to another (say from forestry to global economic development) involves a change in the meaning of sustainability. Whether or not something can continue depends a great deal upon what that something is. But our examples also show how there is a core meaning to sustainability that allows us to extrapolate from one practice or process to another. When we talk about sustainability as a social goal, however, we note some important changes since the Brundtland report brought its definition of sustainable development to prominence in the late 1980s.

When people first began to talk about sustainability in the 1980s, the key challenges revolved around resource depletion, pollution, and preservation of natural variety. Economic development requires energy, but fossil fuels are finite (and potential supply was viewed as more limited in 1987 than it is today). Industrial processes cause pollution, which has harmful effects on human health. Sustainable development was imagined as a process that would reduce pollution dramatically while increasing the efficiency of resource use. But for less industrialized countries, preserving rainforests or vast tracts of undeveloped areas that were havens for wildlife may have seemed in direct conflict with their goal of economic growth. The commitment to sustainable development was, in that context, a commitment to increasing the wealth and welfare of the human population, while also ensuring plenty of room for nature and minimizing environmental damages. By the 2020s, environmental scientists have learned that human societies cannot protect natural variety simply by drawing a line on a map and restricting human access to or use of these protected areas. Succinctly put, the polluting impact of emissions from industrial society is much more widespread than was previously thought. Gases such as carbon dioxide and methane interact with solar radiation to create perturbations in cycles of warming and cooling, wind, and rainfall. The gradual rise in average global temperature associated with climate change is but one of these effects. As people observe change in these cycles, they rethink some key social goals. What they mean by sustainability shifts accordingly. In some cases, people are saying that it's time to move beyond sustainability; the new goal should be resilience.

What is resilience?

Defined narrowly, resilience is a measure of an ecosystem's ability to recover after a very stressful event like a flood or a forest fire. But the word is also used for other types of recovery or rebound. A company able to return to profitability after a major economic setback might be called resilient, and a community that weathers and recovers from an economic or social catastrophe (like losing a major employer or experiencing a pandemic or a destructive episode of racial violence) might also be called resilient. This broad way of thinking about resilience has long been one of the ways that people measure sustainability. A measure of whether a system (e.g., an ecosystem, an organization, a social group) can recover from major stressors is, in an obvious sense, also an indicator of whether that system is sustainable: Can it continue? We ask about resilience again in the context of ecology (in chapter 3), and the answer is more detailed.

What's the difference between sustainability and resilience?

Good question. Sometimes these words are used interchangeably. When people want to mark a difference, a lack of sustainability would usually be associated with a shortfall in key resources or inputs that are needed for a practice or process to continue. Worries about running out of oil or water are seen as challenges for sustainability in just this sense. In contrast, when a process or system of practices is especially vulnerable to disruption, the problem might be described in terms of inadequate resilience. Our approach in this book is to treat the factors associated with resilience as indicators of sustainability. In contrast to a shortage of resources, sustainability as resilience highlights weaknesses in the way systems are organized, like poor coordination of responses or blockages in the flow of information. We give more examples as we move through each chapter. But readers should be aware that the tendency to see resilience and sustainability as two different things may be growing.

Why has sustainability become fashionable? What is it good for?

We will admit that the trend toward sustainability is partly just a matter of doing whatever is currently in vogue: everybody seems to be talking about sustainability, and so there is a bandwagon effect. However, interest in sustainability is also growing for more important reasons. More people are becoming aware of climate change. People want to keep doing many of the things they are doing, and, even more importantly, some practices are essential to our survival: eating food and drinking water, for instance. If there were reasons to think that our access to these essential goods was threatened by climate change (or something else), that would be evidence that our way of life is not as sustainable as everyone would like it to be. Although readers should not lose sight of this bedrock reason for being interested in sustainability, there are, in fact, more subtle and perhaps more interesting ways in which thinking about sustainability has become important.

When someone makes an explicit attempt to gauge how sustainable humanity's current practices are, they're typically led to consider how one process or practice is dependent on and affects many others. That is, they think in terms of systems. Discovering interdependencies and interconnections through systems thinking is useful in itself. It can uncover vulnerabilities but also opportunities that someone might otherwise overlook. Thus, one of the reasons why sustainability has become trendy in fields ranging from business management and urban planning to architecture, farming, and international development is that achieving a better grasp of sustainability can help people achieve a host of other goals: security, profitability, and even social justice. Thinking about sustainability is thus useful as a tool; it can help ensure that the processes and practices people depend upon will continue to be operable and effective in the future. In other words, sustainability is a bridge to systems thinking, and systems thinking is useful for achieving many goals.

What is more, systems thinking can help us understand the way in which many evils recur and persist. In coming to a better understanding of what is needed to make the positive elements in our way of life secure and resilient, people gain tools for understanding why the negatives revisit us with such depressing regularity. A person or group can apply the conceptual, planning, and managerial tools gleaned from analyzing whether processes and practices can continue to processes and practices that they would like to see controlled and ended. In general, then, despite sources of confusion and disagreement associated with the idea of sustainability, integrating measures for sustainability into our thinking increases the capacity to maintain and improve the positive features in ways of life pursued by cultures and social groups around the world. It helps people understand the mechanisms that reproduce unwanted features of social life as well.

In short, once people begin to think about what it is that makes a practice or process more sustainable, they find themselves being drawn into a more powerful way of thinking about a lot of things that are important to them. Although facts about resource consumption and environmental vulnerabilities are crucial aspects of the recent trend toward sustainability, we think that the underlying power of thinking in systems is really what everyone should know about sustainability. So we've organized the rest of the book to help readers get the hang of thinking about how the things that matter to us depend on the continued functioning of both social and natural systems. We start with examples that involve money because we hope they are more familiar. Almost everyone can appreciate how having a constant flow of money coming in is crucial to whether our way of life can continue. We move on to natural systems that continue to function and regenerate themselves with little input or direction from human beings. In both cases, measuring sustainability requires one to understand how each component is embedded within a larger system.

Is pursuing sustainability an individual or social responsibility?

Both. Most people already do at least a little thinking about the sustainability of their own household or way of life. They try not to overspend, and they repair their homes or automobiles long before they find themselves in a crisis. These ordinary types of planning are contributions to sustainability at a personal level. At the same time, it will become obvious that most of the examples we discuss involve larger and more complex social and natural systems. Whether a regional ecosystem or economy is sustainable is a function of the way that many individuals interact. None of us can secure the sustainability of these larger systems all on our own. What people do as individuals adds up, however. A pattern of individual choices can make these larger and more comprehensive social and natural systems unsustainable.

Some readers may be troubled by our claim that pursuing sustainability is a responsibility and our earlier claim that it can be evaluated with data and analysis. In reply we say, "Hey, give us a break! Haven't we been over this before?" We support pursuing sustainability as a more responsible way of thinking about our personal and social options, but this doesn't mean that we have a moral obligation to help any randomly chosen practice or process continue. As we said above, evaluating the sustainability of a practice or process and evaluating whether that practice or process should continue are not unrelated. It would be foolish to pin your hopes on a process that cannot be sustained. Although the tension between these two evaluative processes becomes evident from time to time, we hope that the context in which we discuss any given practice or process will help resolve it. That includes the big, vague system of processes we call "our way of life."

Even if individuals lack the power needed to bring the behavior of their fellow citizens in line, they can help to make their societies more sustainable by acting in ways that promote the continued functioning of key social and ecological systems. Anyone can contribute in three ways. First, since consequences add up, what people do as individuals contributes directly to the overall maintenance of social, economic, and environmental integrity. Second, living a more sustainable lifestyle models a key form of civic virtue for all of us living in the twenty-first century. By reflecting on the way our actions interact with the behavior of others, one exhibits the qualities of responsible citizenship needed to achieve greater sustainability at the social level. Finally, individuals can act through the political process to encourage businesses, policymakers, and other leaders to make decisions that improve sustainability at a social level. The last chapter discusses some specific things that individuals can do to promote sustainability and especially some important questions to ask about those things.

How can I use this book?

Books in Oxford University Press's What Everyone Needs to Know® series are intended to give the reader an introduction to essential ideas in technically complex areas of subject matter. They pose a series of questions that anyone trying to get a handle on the main topic might ask. Our answers explain key ideas and offer illustrations and examples to show how these ideas apply to situations that any reader will know. We have organized these questions so that a reader who works through each of them in order will gradually build up more and more familiarity with the systems way of thinking that is at the heart of sustainability. Nonetheless, readers should feel free to skip ahead to the questions that most interest them.

We start in chapter 2 by looking at the way concepts of sustainability are used in running a business because we presume that people will have some fluency in understanding what it takes for a firm to survive in a competitive economy. Chapter 3 illustrates how similar ideas are applied in developing an understanding of ecosystem processes. Chapter 4 continues the environmental theme by discussing some key ways sustainability thinking has been used as a response to pollution, resource depletion, and other threats to environmental quality. Chapter 5 introduces readers to basic ideas in economic development and explains how the Brundtland Commission's approach to global economic development came to dominate thinking on sustainability in general. The Brundtland Commission was motivated to rethink global development in the context of limits to growth and framed its approach in terms of being fair to both future generations and the impoverished people of our own generation. This leads into the topic of chapter 6: sustainability and social justice. As we already noted, sustainability is inherently bound up with our ideas of good and bad, of fairness and unfairness, and of progress and degeneration. We try not to preach, but chapter 6 provides an overview of the way that ethics and values shape discussions of sustainability. This is followed by a chapter on sustainable governance. In chapter 7, we frame our discussion around two queries: What factors influence the sustainability of governance systems? And how do governance processes affect sustainability more generally? In chapter 8, other domains of society—the arts, religion, and especially science—are discussed in light of the sustainability ideas developed in earlier chapters. The book concludes with a chapter considering what everyone needs to ask about promoting sustainability.

2

SUSTAINABILITY
AND BUSINESS

Why start with business?

We believe that most readers have an intuitive grasp of what
it takes to run a business, even if they have never actually
done so. The mindset of the business operator is geared toward a basic understanding of sustainability that can serve as
a model for other applications of the concept. We build on this
throughout the rest of the book. People commonly say something like this: for a business to be sustainable, it has to be
profitable. There are reasons to qualify this blanket statement,
and we will get to them soon. The basic idea is that whether a
company is manufacturing widgets or selling them, whether
it is extracting natural resources or delivering services, all
firms must take in enough money to cover their costs or they
will eventually go broke. This idea can be a starting point for
making sense of sustainability in many non-business domains.

At the same time, many people who are deeply committed
to sustainability are wary of the business mindset, and others
are actively hostile to it. Again, there are reasons for suspicion
and even enmity toward for-profit operations, and we will get
to them eventually. We ask readers who come to our book with
a predisposition to be cautious toward or resistant to the business mindset to cut us a little slack here. Obtaining a firm understanding of the business operator's perspective can help us

think about other dimensions of sustainability, so it is worth taking pains to make sure all our readers are working with the same idea.

What does profitable mean?

Technically, a firm's profit is equal to its total revenues minus its total expenses. In other words, a profitable business is one that is able to cover all its costs and still have something left over. In a small business, the profit goes to the owner, and they may reinvest it in the business or treat it as additional income beyond their salary. For a large company with external stockholders, profits may be reinvested in the business, or they may be paid to stockholders in the form of dividends. In both cases, profitable periods help businesses weather periods of financial losses. Over the long term, in the absence of profits, businesses and investors take their time and money and invest them elsewhere.

Many small businesses exist where the owners/operators do not pay themselves a salary commensurate with what they could earn for comparable work elsewhere. Such a firm can continue to exist for a long time if the owner/operator is willing to put up with that level of income. Strictly speaking, students of business and economics would be taught that such businesses are not maximizing profits when resources (in this case, the owner's labor) are not being used in a way that maximizes the return to their employment. Yet these businesses may be sustainable in the sense that they can keep on going. Thus, the idea of a *going concern* captures the idea of sustainability we are developing here better than the technical concept of profit as it is taught in business and economics.

This is, of course, a very simple explanation of how businesses manage their finances. However, it enables us to address one of the reasons for wariness and hostility toward businesses and profits among some sustainability advocates. Because being profitable technically means earning more revenue than is needed to cover all operating expenses, critics argue that the pursuit of profits causes businesses to overlook other important goals in the interest of financial gain. Those other important goals, they argue, include things like environmental protection and fair labor practices. No doubt, examples of such myopic business practices can be found. Yet digging a little deeper into this argument uncovers a broader philosophical concern—that economic activity, including profitable businesses, feeds a level of consumption that is unsustainable. Viewed as a group, humans produce more than they need and are depleting the earth's resources. This is an idea we give more consideration in our chapter on sustainable development. Nevertheless, it doesn't detract from our basic point that businesses must be profitable to be sustainable, to remain a going concern.

Why does the business mindset see sustainability in terms of profit?

The bottom line is an expression that refers to an accountant's spreadsheet tallying a firm's revenues and expenses. Operating a business involves spending money on wages, salaries, materials, and other operating expenses and collecting payments through sales or other sources of revenue. If spending exceeds revenues for very long, sooner or later the business will reach a point where it cannot continue to pay for the materials and labor it needs to conduct the activity that generates income. In a typical scenario, the business is unable to meet obligations to pay employees for work done or to pay creditors that have delivered supplies and extended loans, or are expecting the payment of rent. In a simple bankruptcy, the firm ceases to exist. A court must decide how to distribute the remaining assets (bank accounts, unsold merchandise, or unused supplies) to the people who are expecting to be paid.

Admittedly, things are not so simple in modern economies, but we are not here to tell you what everyone needs to know about managing a business. The point is to draw upon an easily graspable aspect of business to illustrate our basic definition of sustainability: a measure of whether (or to what extent) a process or practice can continue. In this case, the process or practice is *doing business*: making and selling a given type of good or service. The company or firm itself is not sustainable unless it takes in enough money to cover its expenses. This is the practical wisdom behind the idea that profitability is the measure for sustainability in a commercial enterprise, company, or corporation.

Notice that this basic idea of balancing revenues and expenses can be extended to organizations that do not actually define success in terms of profit. Ordinary household management works this way, and if someone spends more than they earn for too long, they will find themselves in bankruptcy. The same is true for nonprofit organizations such as churches, clubs, or civic groups.

Are all businesses committed to sustainability then?

No. Our simple example might make it seem like the sustainability of the firm would make every business operator keen on staying in business, but the example is too simple to capture the full meaning of profitable business practice. Bankruptcy laws allow firms to shed their debts and continue to operate time after time, showing why the real world of business is more complex. Plenty of people participate in business activity with no thought of sustainability even in the limited sense we've explained so far. We increasingly hear about the business world as a sphere where sharks and speculators search for "the killer app," "the home run," or "the new new thing." They are less focused on long-run profitability than on finding exceptional investment opportunities with large short-run returns. They might even liquidate a business and sell its assets—the very opposite of sustainability in the sense we are discussing. The story is complicated when it comes to venture capital investors and innovators, but, again, we are not trying to give a full account of the business mentality. To stay on track here, we are using the idea of a profitable business to motivate a more general understanding of sustainability.

How does a business manager think about sustainability?

If one focuses on those going concerns that do hope to continue doing business on a day-to-day basis, the measure of sustainability is taking in more than you pay out. But a shrewd manager recognizes that both *taking in* (i.e., revenues) and *paying out* (i.e., expenditures) involve interactions with other firms that are also watching the bottom line. These larger economic realities themselves exhibit aspects of sustainability. A business operator has more direct control over expenditures, and managing them often occupies much of their attention. As noted, businesses have to pay workers and buy supplies. They also have to pay rent, taxes, and fees (such as interest on loans or membership in organizations). To the extent that they can lower any of these costs without also reducing what they take in, the bottom line looks better to them. A business operator will often describe activities that reduce their costs as increasing the efficiency of their operation.

Not everyone thinks of efficiency in just this way, but it is a good starting point for understanding why increases in efficiency are often seen to be the route to improving sustainability. One particularly significant dimension of cost efficiency is tied to labor. The industrial revolution led to dramatic increases in the efficiency of labor, especially in manufacturing. What this meant to factory owners was that spending on new machinery would be more than offset by the amount of product they could churn out for the same amount of wages and salary that they were paying to their workers. They would therefore reduce the amount of expenditure (e.g., wages) per unit of product. If they could sell whatever they were making for the same price they had always been getting (i.e., keep revenues stable), they would make more money over the long run. This would thus enhance the sustainability of the business.

There are two takeaway points here. First, it is natural for a business operator to focus on expenses, because that's where they have the most leverage to affect the ratio of expenditures and revenues (i.e., the bottom line). Unless they can envision an increase in the cost-effectiveness of their operation, they will resist anything that leads to an increase in costs (like taxes or a minimum-wage rule) for the simple reason that it challenges the sustainability of a going concern. They will also be very quick to jump on changes that reduce their costs. Since any one business is probably buying supplies from another business, increases in the cost-effectiveness for other firms can also improve a business's bottom line. The second point follows. Although this story rapidly becomes very complicated (leading to the emergence of economics and business administration as domains of specialized knowledge), the overall health of the economy plays a huge role in any individual firm's ability to sell goods and services. If workers are themselves making money (i.e., the economy is strong), more people will want their widgets, strengthening the sustainability of the firm. And, of course, the reverse is true. Shrewd business managers therefore see that there are links between the sustainability of their firm and the sustainability of the economy as a whole.

This is not the whole story, of course. Non-economic events can have an enormous impact on a firm's business activity. A spike in oil prices after Iraq's invasion of Kuwait in 1990 affected businesses (like airlines) whose profitability is linked to fuel prices. In 2011, a tsunami hit the Pacific coast of Japan, causing a large number of deaths and a major accident at the Fukushima Daiichi nuclear power plant. It triggered a disruption in economic activity that was felt across the world as supply chains for manufacturers in other countries were interrupted. In 2020, schools were closed and bars and restaurants in many places were ordered to close in response to the

COVID-19 pandemic. It may be impossible to remain profitable over the short run after shocks like these, but in planning for a rapid return to profitability, managers incorporate sustainability thinking when they think about resilience. Many companies are developing plans for how to be more resilient in the face of climate change and associated changes in the availability of natural resources and environmental amenities that are critical to their businesses. We discuss this point more shortly.

How does business sustainability relate to the economy as a whole?

Business operators understand that it is easier to sell products and services when people have money to spend, so the sustainability of a business is enhanced when the overall economy is healthy. This is usually associated with economic growth, and this leads to the question: When is economic growth sustainable? This is a topic we cover in chapter 5. But this relationship between the sustainability of a given business enterprise and the sustainability of the larger economy illustrates one of the basic ideas in sustainability thinking: hierarchy. Systems exist within systems. As one moves from component system to the larger, more comprehensive system, one is moving to a higher level in the hierarchy. If you think about individual businesses as subsystems that exist within the larger economic system, you see two important things. First, more comprehensive systems—in this case the economy as a whole—are composed of smaller units that are themselves systems—in this case individual firms. Each firm can be thought of as a subsystem of the larger economic system. Good times, bad times: adequate revenue is essential to any going concern. Second, the sustainability of any given firm is independent of the overall sustainability of the economy. Some firms may do well when times are tough, after all. More generally, the criteria for the sustainability of a subsystem (in this case the profitability of an individual firm) are just different from the criteria for the sustainability of systems at a higher level in the hierarchy (that is, the economy itself).

Collectively, the sustainability of individual firms does affect the sustainability of the larger whole; when the number of failing firms reaches a tipping point, the entire economy will take a dive. Correlatively, when the larger system is faltering, it tends to threaten the sustainability of the subsystems. When the economy is bad, it is harder for individual firms to make a go of it. Thus, although you have to evaluate the health of the economy in different terms than you use to evaluate the financial sustainability of any individual firm, the criteria are related. Getting clear on the relationship between a subsystem and a larger system is one of the difficult technical problems in evaluating sustainability. The principle of hierarchy is a basic concept for sustainability thinking, and readers will find it at work in every chapter of this book.

But businesses have always understood the need for profit. Why is sustainability different?

Our starting point was simply to illustrate a very basic application of the general concept of sustainability. Connecting the sustainability of a given firm or product line to profit is just common sense. But it is also true that starting in the 1990s businesses began to think about sustainability in a somewhat different way, one that often paid much closer heed to the environmental impacts of doing business. This new way of thinking was based on an earlier trend in business management, corporate social responsibility, or CSR. CSR was initially understood as managerial practices reflecting the view that firms should conduct business in a way that improves social well-being, beyond merely producing desired goods and services. As with sustainability, CSR involved considerable flexibility in terms of what actions a firm might take to be socially responsible. In some cases, CSR meant meeting and even going beyond the minimum requirements of environmental and other laws under which a firm might operate. In other instances, CSR translated into more direct involvement in social causes or community activities. For example, a large corporation might make a high-profile donation to a well-known charity such as an art museum or an NGO working to find the cure for a deadly disease. A smaller business might sponsor after-school activities, donate uniforms to a youth-league baseball team, or provide incentives for employees to volunteer in their communities. Businesses engaged in CSR activities for two reasons. One was that this kind of activity gave the company *social permission to operate*—implicit authorization, endorsement, or consent for them to conduct business in a given community. The other was the owners' desires to give back to the community in much the way that anyone might make a charitable gift or undertake volunteer activities.

Social permission to operate and CSR opened the door to ideas that became important for sustainability within the business community. The rest of this chapter takes the reader far beyond the common-sense notion in which sustainability can be tied to a firm's need to take in more than it spends. However, from a business management perspective, all the ideas we discuss are compatible with the need for profit. Depending on circumstances, attending to these concepts may increase a firm's profitability or even be a necessary condition for continuing to function as a going concern. One of these concepts is social capital.

What is social capital?

The basic idea here is that when people like and trust you, their confidence in you enables you to work effectively. On the other hand, if you are distrusted, people will be constantly looking over your shoulder to make sure every little thing is done precisely to code (that is, strictly by the rules). Put another way, no one will buy from you if they don't trust you. A company that lacks social capital may be able to make money, especially if the product they are making doesn't need to be sold in their local community, but they may find it difficult to attract good employees, or they may be beset with other annoyances. When neighbors or consumers lose trust, firms may find themselves facing nuisance complaints or, worse still, fighting lawsuits. This sense of trust and confidence that a business will obey the laws and will look out for interests of neighbors and workers is referred to as social capital. (We put off the answer to why this is called capital until chapter 5.)

Managers started to link social capital to CSR and to think of CSR in terms of the community's level of confidence in a company during the 1970s and 1980s. But as they started to think this way, it was obvious that social responsibility in this sense meant much more than giving to charity. Noxious odors, loud noises, and the traffic congestion associated with trucks and shift changes are likely to generate a complaint from a company's neighbors. Responding to these complaints can get expensive. As CSR became more and more attentive to a company's social permission to operate, managers started to describe the need to address these problems as part of their ethical responsibilities. At least for manufacturers, CSR began to include the responsibility to take care of their local environment as part of a business's need to build social capital.

How does social capital relate to sustainability?

If you focus on the business operator's perspective, the answer to this question is that business schools and commercial firms started to use the word sustainability to name aspects of management that were focused on building trust and goodwill (and thereby reducing costs). This was totally compatible with the overall idea that a going concern must keep its eye on the bottom line. But in an indirect fashion, building social capital involved a set of activities connected to the firm's main business activities—making salable products and delivering billable services. Consumers have driven many decisions about building social capital in this context by asking for products and enterprises that are more attentive to human and environmental health. Some companies created departments or divisions that were dedicated to sustainability in just this sense, and large firms created positions such as director or vice president of sustainability to oversee these activities.

How does social capital relate to the environment?

From the perspective of someone outside the company, many of the things that firms did in this new management space related to the environment. They managed pollution, noise, and impacts on the social environment that could otherwise cost them money in the future. And similarly, cleaning things up or providing amenities to neighbors in the form of parks or wetlands generated goodwill for companies at the same time that it was perceived as an environmentally oriented activity by outsiders. This created a loose but plausible connection between what businesses did to increase their social capital and what the public at large associated with growing environmental consciousness.

Moving back to the company's own perspective, sustainability managers within many firms began to discover that they could make choices in their business operations that could achieve environmental benefits above and beyond the environmental protections required by law without imposing significant costs on their own operations or even reducing some operating costs. For example, in the late 1980s Dow Corporation initiated its Waste Reduction Always Pays (WRAP) program with a goal of reducing its waste stream and thereby reducing waste treatment and disposal costs. It included a program for rewarding employees who came up with ways to reduce waste—either by reducing throughput or by using waste materials to produce another product. More recently, Walmart began requiring its suppliers to measure and report the energy and materials used in making their products. This allows Walmart to base their choice of which products they will sell in part on how they affect environmental quality and total resource consumption. This choice became a part of Walmart's sustainability strategy, allowing Walmart to build social capital by contributing to environmental quality at no significant cost to themselves.[1]

We should also point out that the association between social capital and the environment did not always have a positive connotation. People who were skeptical about environmentalism and environmental consciousness perceived these activities as being forced upon companies against their will, making their own business activity more costly than that of competitors (usually in other countries) who did not need to undertake expenses that (in their view) were not delivering any real health or well-being benefits. This might, for example, have been the case for companies hoping that their products would be sold in Walmart stores. They did incur increased costs in supplying information on energy and material use to Walmart and may have also incurred expenses in changing their own manufacturing processes to get their products onto Walmart's shelves. If they did not believe that real benefits to public health or environmental quality were being realized by these changes, they probably saw sustainability as a waste of money.

What is the triple bottom line?

This is the phrase that business managers use to promote their approach to sustainability. The traditional profit-oriented bottom line is being supplemented by the need to build social capital on the one hand and promote environmental quality on the other. If you skipped ahead to this question, going back to "What is social capital?" will help you make sense of this answer. Triple bottom line is a phrase credited to business author John Elkington, who also coined the expression "people, planet, and profit" to communicate how sustainability needs to address the social, environmental, and economic bottom lines. Up to this point, we have intended to show how a simple notion of watching the bottom line was expanded to include social and environmental dimensions through an appreciation of the way that public trust and goodwill are critical for a going concern's ability to operate profitably. Elkington was one of several authors, including Paul Hawken and William McDonough, who popularized and spread these notions in the business community. Recently, Elkington has expressed disappointment that the triple bottom line concept became a simple accounting tool that fostered a trade-off mentality and failed to stimulate a broader discussion about capitalism and its future.

We've also hinted at some of the reasons why others have been skeptical of the triple bottom line. A manager working within a large corporation may not have an incentive to take long-term payoffs into consideration. Positive social or environmental outcomes could occur years, decades, and even centuries in the future. Businesses find it difficult to reward managers for benefits to future generations, even when they can anticipate them. In addition, business decisions are made within a legal environment that limits a profit-oriented manager's options. Too much cooperation among competing firms can provoke antitrust measures. The decisions made by competitors can also limit a business's options. Managers might want to pursue more sustainable practices but feel constrained by aspects of the larger system that are beyond their control. All these considerations should lead us to recognize that more comprehensive systems influence the incentives we have simplistically tied to profitability. In the balance of this chapter, we consider some questions that might be raised by people who harbor suspicions about whether a business mentality can truly achieve sustainability.

What is greenwashing?

Whitewashing refers to things that people do to conceal ugly, illegal, and immoral practices behind a veneer or façade while doing nothing to actually correct them. Greenwashing occurs when the façade is painted in environmental hues. Many critics of the triple bottom line have argued that companies are claiming to have cleaned up their act with respect to pollution, resource consumption, and other forms of harmful environmental impact but that it is just propaganda.

Greenwashing occurs in advertising, public relations, and product claims. Advertisements using images of nature or claiming products are natural are common examples because there is no legal or commonly agreed-upon criterion for naturalness. A company's public relations greenwashing may be harder to spot. Public relations greenwashing covers everything from posting in discussion forums on the Internet to planting stories in the newspaper. However, product claims may be the most hotly contested area for greenwashing. In some cases, there is little difference between advertising and a label appearing on a product, such as new or improved. However, when labels or advertising make specific claims about a product's health benefits or environmental impact, they are supposed to be true, and companies can be prosecuted if they are not. Of course, government resources for monitoring product claims are limited (more on this in chapter 7). As a result, new methods for controlling greenwashing have emerged.

In no small part because of consumers' desire to understand more about environmental protection and social responsibility claims made by companies, a vast array of certification and labeling programs have evolved to help provide consumer confidence in such claims. These programs generally involve a third-party entity that reviews processes and practices and authorizes companies to label their products as "environmentally responsible," "organic," "fair trade," and even "sustainable." A few examples of such programs are described in chapter 7. Has third-party certification eliminated greenwashing? The answer is clearly no, but it has both reduced it and given consumers information they otherwise would lack.

Certification programs have not ended greenwashing because it is possible to develop relatively meaningless certifications, as well as certifications that become misleading. Scientists dispute the environmental and food safety benefits of certifying foods as GMO-free (to signify that they do not contain ingredients that have been genetically modified). Organic food standards do imply that the most toxic chemicals have not been used, but they include neither protecting the welfare of animals nor fair treatment of workers. In addition, the chance that producers will cheat remains. *Quis custodiet ipsos custodes?*—Who will guard the guardians?—is a phrase that goes back to Roman times. Nevertheless, certification schemes have put a check on greenwashing, and the fact that people will pay more for certified products tells the business community how much the public values sustainability.

How do businesses fail to address social responsibility?

We noted that the business sector's approach to sustainability grew out of CSR, the corporate social responsibility movement. Our treatment stressed social capital and moved quickly to environmental issues such as pollution and aesthetic amenities like parks or wetlands. But CSR itself grew out of a much longer history that was focused on the harmful social impacts of business activity. It started with the simple fact that a company can always make its bottom line look better by reducing costs, and one big way to do that is to pay less for labor. Managers are taught to gauge the level of compensation to workers based on the going rate for similar work, or colloquially "whatever the market will bear." Workers, however, want to be paid as much as possible and quite probably have expectations about what level of compensation is fair. Paying workers less than what they think is fair makes them unhappy, and paying more than the going rate makes management unhappy. This basic conflict will not be a surprise to any of our readers.

The twentieth century saw a lengthy and complex history of strife and negotiation between labor and management. The rights of workers to organize into trade unions was a key component of this history, with some industries and some geographical areas seeing a high degree of unionization, while elsewhere business operators were successful in resisting unionization. Contrasting views on this issue continue to divide political parties, and these same contrasting views inform debates on sustainability. Put succinctly, some people would argue that sustainability includes or implies the worker's side of this longstanding social debate. A company that opposes unions or some form of collective bargaining is, in this view, failing to address the social side of sustainability. Interestingly, those who feel that strengthening workers' rights has a deleterious effect on the economy hardly ever relate their views to the idea of sustainability. Yet the legacy costs associated with union contracts in the auto industry likely played a significant role in the Chrysler and General Motors bankruptcies during the Great Recession of 2008 and 2009. Meeting worker demands was a sustainability issue for these companies.

This link between social responsibility and the rights of economically weak or disadvantaged parties can be broadened well beyond unionization and the right to organize. Some would see redress of racial, ethnic, and gender disparities as aspects of sustainability; they emphasize how injustice is a systemic feature of the economy. We dive into these broader concerns in later chapters. In keeping with this chapter's focus on business, we note that there are many ways in which particular businesses and the business community in general have been criticized for unequal treatment of people based on their race, gender, or sexual orientation. Thus, promoting fair treatment of individuals and rectifying the relative lack or underpayment of women or minorities in the business world is seen as a component of a business's social responsibility. If one tends to think that irresponsible actions can't be sustainable—and we have noted already that some people think that way—then unfair treatment of employees is, in this view, unsustainable.

All of these themes can be seen in the early literature of CSR, and they have subsequently been carried over into the push for sustainability. Collectively, these questions about labor rights, economic inequality, and the rights of marginalized or disempowered groups point toward questions of social justice. When these left-leaning perspectives on social justice are advanced under the banner of sustainability, the concept of sustainability itself becomes politicized. This likely explains why conservative columnist George Will devoted a 2015 column to a scathing critique of sustainability programs at American college campuses.[2] We return to some of these political questions in chapter 7, but for now we maintain the business focus.

How is social justice related to sustainability in business?

We should probably start by admitting that people who are irresolutely hostile to the very idea of capitalism quite possibly think that profit seeking is simply incompatible with any globally sustainable society. From the perspective of CSR, however, the answer goes back to social capital. On the one hand, CSR attempted to more faithfully realize the socially positive aspects of competitive profit seeking. Managers were encouraged to be more cognizant of ways that economizing on labor costs and shifting costs to others create hardships within the communities in which they are operating. On the other hand, CSR heightened managers' understanding of the ways in which building trust, confidence, and goodwill among those who are not directly involved in a company's activity actually can make positive contributions to a company's bottom line. Over time, this more enlightened and comprehensive approach to profitability has prevailed in most business schools. The Harvard Business School has ended its programs on corporate social responsibility and incorporated these ideas into an approach that emphasizes building shared value.

Current trends in management thinking thus see sustainability as one component of a bigger picture in which sophisticated approaches to profit seeking take account of the fact that
businesses operate in a complex social environment. Stated
simply, good managers avoid costly lawsuits and labor actions by thinking strategically about all the myriad ways in
which their operations can incur avoidable costs. A going concern manages the risk of unplanned-for expenditures caused
by ill will by cultivating a good relationship with employees
and with others in society. For such entities, social justice is
seen in terms of fair dealing and anticipating the risks arising
from social ill will. The traditional management perspective
on increasing shareholder value through profitability inter-
prets these activities as building social capital. From this perspective, a company that is building its social capital is also
attending to social justice.

But the traditional management approach is incomplete so
long as its focus remains on those aspects of social justice that
are closely tied to companies' own activities and remain within
managerial control. Lawsuits and labor actions may be aimed
directly at a specific company, but more general social disruptions can also imperil a firm's ability to operate successfully.
Rioting in the streets is seldom good for business. When social injustice spills over into disruptive forms of social protest,
or even revolution, the overall social and political environment may expose firms both to physical losses (shop owners
who must replace broken windows or looted inventory) and
to downturns in normal business activity. It is thus possible
to see extreme forms of social injustice as inimical to business
sustainability, even when the injustices are not directly related
to the policies or activities of the specific firm in question. This
creates a reason for business operators to take an interest in political activities that affect the social environment in which they
operate. We take up some of these themes in later chapters.
Elkington's "triple bottom line" approach urged managers to
be more attentive to such matters, but it is worth pointing out

that these more expansive or global ways in which social justice becomes relevant for business sustainability are seldom incorporated into the strategic thinking of business sustainability. This also points us to other ways businesses actually depend on the stability or sustainability of more comprehensive systems—economic systems or systems of government—when they plan for their own sustainability.

How does business depend on the environment?

Ecosystem services is a name for all the ways in which humans rely upon and benefit from the things that natural systems provide. A healthy environment provides most businesses with necessary services like clean air and water. Accounting for the importance of ecosystem services to the bottom line—or to long-term sustainability—is an increasingly common part of how businesses manage their operations. Businesses rely upon ecosystem services if their production processes require inputs from the natural environment or if functioning ecosystems enable them to operate at lower cost than they otherwise would. We discuss ecosystem services in more detail in chapter 4.

For now, consider how important a readily available supply of clean water is to so many types of businesses, not just for employees to drink but for inclusion in products (think bottled water, beer, and soft drinks) and for production and cleaning processes (think agricultural irrigation, food processing, and automobile manufacturing). In 2014, a General Motors manufacturing plant in Flint, Michigan, discovered that automobile parts they were producing were being corroded by the water coming to the plant from Flint's public water source. Flint was drawing from the Flint River, and chloride levels in the water were so high that the metal parts were being damaged. Managers at the plant tried a number of cumbersome and costly methods to treat the water and reduce chloride levels. Ultimately, they disconnected the plant from the Flint water system; the plant was able to connect to the water supply of a bordering community that drew its water from a different source.[3] This example makes it clear how businesses' interests can be threatened when critical ecosystem services are lost.

Other types of businesses that depend upon ecosystem services have experienced challenges when critical ecosystem services were threatened or lost. For example, restaurants and other companies with seafood at the core of their business activity have struggled as key fish and shellfish stocks have dwindled. Extreme flooding events, expected to become more common with global climate change, have destroyed business infrastructure and opportunities, product inventories, and supply chains. Corporate Eco Forum is an organization of global companies that have articulated commitment to environmental protection as part of their business strategies. Viewing protection of ecosystem services as a business imperative, these companies recognize that protecting ecosystem services reduces risks, cuts costs, and enables them to enhance their brands and achieve growth in their businesses. Restaurant chains that specialize in seafood have invested in rebuilding threatened fisheries and implemented requirements for sustainable fishing practices for their suppliers. Companies dependent upon wood and cotton fibers for production of their products are investing in alternatives—like flax, bamboo, and byproducts of agricultural products such as wheat straw—to reduce pressures on forest resources and the environmental impacts of cotton production.

How does business affect the environment?

We don't want to get too far ahead of ourselves here. *All* of us affect the environment by using energy to heat our homes and run our vehicles. People put waste into the environment when they dispose of packaging and uneaten food. They draw upon water supplies to bathe and wash clothes. Our use of toxic substances (sometimes concealed in packaging or exhaust gases) pollutes air and water, and when our consumption releases greenhouse gases (as when people use fossil fuels), it contributes to climate change. Businesses do all these things, too and in this sense we are all in this together.

However, businesses typically do all these things on a larger scale than the average household. They also make key decisions that have large impacts on the way individual consumption affects the environment, so it is worth pulling this theme out before closing our discussion of sustainability and business. The larger point, of course, is that businesses depend on their ability to use natural resources and on the various services that ecosystems provide in order to manufacture goods and provide jobs. So their ability to be sustainable, much less profitable, depends upon the continuing availability of those resources and ecosystem services at the same time that business activities are undercutting their ability to depend on them in the future. This profound lesson lies at the core of what everyone needs to know about sustainability. The smart, responsible managers would look beyond squeezing the maximum profit from their operations and consider how they might change their activities so that the next generation of managers can watch the bottom line as well. But they do not control the larger system that often rewards short-term decisions over long-term sustainability, and this limits their ability (and willingness) to make the smart, responsible choice.

What is the take-home message?

When people with a business background associate sustainability with profitability, they are drawing on sound intuition. The flows of expenditures and revenues have to balance out with revenues at least equal to (and hopefully exceeding) expenditures on the bottom line. That's what profitability is, and a firm that allows expenditures to exceed receipts will not be a going concern (it will not be in business for long). That is, it will not be financially sustainable. Faculty at business schools began to introduce systems thinking into management curricula in the 1960s. Business operators have now learned that their own profitability depends on the health of the general business environment (i.e., the economy) in which they operate: they understand the principle of hierarchy. Some businesspeople think more broadly and recognize that social interactions occur within a biological environment that can also affect their profitability. Understanding of this hierarchy may have increased after the COVID-19 outbreak of 2020. This recognition leads them to consider how their firms will not be sustainable unless both social and biological aspects of that environment function in ways that are not overly hostile to their business activity. So thinking in business terms can lead managers to take account of many of the larger features that impact sustainability.

At the same time, there is much more to the story.

3

SUSTAINABILITY
AND ECOLOGY

What is ecology?

The current Wikipedia entry for "ecology" defines it as "the branch of science that studies the relationship between organisms and their environment." But when Marvin Gaye's record "Mercy Mercy Me (The Ecology)" came out in 1971, he was not talking about a branch of science. When he sang "Things ain't like they used to be / Where did all the blue skies go?" he was talking about the state or condition of the natural world. He was also saying that the condition was not what it should be. This sentiment is a good starting point for a chapter on sustainability and ecology; it conveys the sense that something about the current condition of the natural world is not sustainable.

But going back to something we said in chapter 1, the sentiment expressed by Marvin Gaye is ambiguous in a manner that is common in discussions of sustainability: it is both descriptive and prescriptive. It's saying that certain natural processes are at risk, under threat, or reaching a point at which they will collapse and be unable to continue. But Gaye was also saying that the current condition of the natural world should not be allowed to continue. Laying stress on the descriptive side leads right into the branch of science that studies the relationship between organisms and their environment. As a science, ecology does study the natural processes that support living organisms (including but not exclusively human beings). It can provide insight into the ways they are at risk and possibly unable to continue. When the population of organisms being studied is us, it is a very short step to the ethical judgment that these processes *should* continue. The other side of this ambiguity is making a much broader claim: certain states of the world (the way it used to be, in Marvin Gaye's song) are better than the way it is now. And this is a claim that cannot be supported by ecological science alone.

In this chapter, we mostly focus on the way that ecological science helps us derive a measure of whether natural processes can continue, and correlatively of the things that put their continuation at risk. We defer most of the questions about the way that the sustainability of human practices depends on natural processes until the next chapter, "Sustainability and Environmental Quality." As the sustainability of a business— a going concern—can be understood in terms of a system of practices that must happen for it to continue from one day to the next, living organisms interact with one another and with soils, water, and other elements in systematic patterns. Ecological science can provide partial answers to questions about whether and under what conditions these patterns continue and when they dissipate or undergo radical change. The question of when patterns should continue is not a strictly scientific one, but it is a question that at least some people think can be addressed apart from the use that human beings make of natural processes.

How do ecologists approach sustainability?

Ecologists developed a body of theory around ideas that are closely related to sustainability, but they did not always use that word. Change in the size or geographical range of a population provides an example. A group of organisms that interbreed and occupy a given region makes up a population. The number of individuals in the population will grow or shrink based on the availability of food, the rate at which they reproduce, and their longevity. Longevity is partly determined by organisms' physiology, but organisms' survival is also affected by whether the organisms must contend with predators, disease agents, and natural events such as drought, freezing, or floods. The study of how these factors interact yields a measure of whether the population can continue to survive in the given region over time. If the rate of births in a population exceeds the death rate, the population will grow. However, as populations grow, more individuals are drawing upon the food and water resources in the region. At some point, they will exceed the supply of resources, and the death rate will mount. The region's ability to support a population is called its *carrying capacity*. Population growth beyond the region's carrying capacity is unsustainable.

In nature, of course, unsustainable rates of population growth do not occur for long. In one scenario, populations that exceed carrying capacity crash when they have exceeded their food supply, reverting to a much smaller size. Growth may begin again, leading to oscillations in the size of the population. In another more typical scenario, another species emerges to control the population. In a deer-wolf ecology, wolves hunt the deer, bringing the death rate into balance with the birth rate, and narrowing the oscillation in the size of the deer population. In this oversimplified model, predators (the wolves) and prey (the deer) exist in a symbiotic relationship that creates a stable system. It is the system comprising wolves and deer acting as predator and prey that is sustainable. Of course, a region will contain populations of many different species of plants and animals, making the total array of predator-prey relationships quite complex. While the stability of a population as controlled by predators and the underlying resource base provides one example of the way that ecologists might approach sustainability, the key idea is the way that all of these species (together with abiotic elements) create a complex ecosystem.

What is an ecosystem?

An ecosystem is the community of living (biotic) organisms from different species that interact within a region together with the nonliving (abiotic) components that support life processes. The key processes that connect organisms and abiotic elements are nutrient cycles (discussed below) and energy flows. All living things consume energy through metabolic processes that combine organic material from other organisms with water and other abiotic elements (gases from the atmosphere, minerals from the soil) to produce their own bodily tissues and power their growth. Sunlight is the original source of energy, which is transformed by plants through photosynthetic capture of carbon dioxide from the atmosphere and release of oxygen into it. Animals derive energy by eating plants (as well as each other) and emit carbon dioxide as a waste product. The balance between oxygen and carbon dioxide production is just one example of the way that ecosystems balance relationships that serve as the basis for system unity and reproducibility over time.

Since this is not a book on what everyone needs to know about ecosystems, we are trying not to ramble on. A very simplified discussion illustrates how balanced relationships in ecosystems provide the basis for ecological sustainability. The eventual shape and composition of a given ecosystem is itself a balance of factors such as climate and the region's geophysical endowment (e.g., oceans, fresh water, soils, mineral deposits) and biotic components (e.g., different species) and the respective relationships that develop among them. While predator-prey relationships are important, other factors such as shading and evapotranspiration can play a major role in regulating the flow of energy through the ecosystem.

While the notion of balance implies stability, ecosystems are in fact always undergoing change. One way to think of this is through the analogy of your own body, which is constantly replenishing and rebuilding itself from the food that you eat.

Although ecosystems do not age in the way that our bodies age, they do accumulate small changes among the organisms that comprise them over time. For example, forests undergo a process of succession during which dominant tree species shift. As another example, a predatory species may shift from eating one species of prey to another as the old one becomes scarce.

What is more, dramatic events such as a wildfire, a flood, or a drought can create larger-scale disturbances in ecosystems. In many cases, the systemic character of an ecosystem can rebound from such disturbances, a phenomenon known as *resilience*. Ecosystems that are in the process of recovering from some past disturbance will exhibit larger oscillations in the population of organisms comprising them. Such disturbances also create opportunities for once small populations to expand, leading to significantly different configurations of the ecosystem than might have existed before the disturbance. All of this shows that sustainability in ecosystems should not be taken to mean rigid stability or lack of evolutionary change. In an important sense, the stability of an ecosystem has less to do with specific types of organisms than with the way different populations of organisms create functional relationships (such as maintaining energy flows) with each other and with abiotic elements.

Nevertheless, species matter. Ecosystems that share similar geological characteristics can differ markedly in the ways species interact. While the resource endowment in a region is strongly shaped by geology and climate, the way in which resources become available to support life processes is determined by factors that emerge within the interaction of organisms: decomposition of organic matter, competition for soil nutrients and water, or factors like shading that affect the uptake of solar energy. Entry of new species can cause substantial shifts in ecosystem function, whether caused by natural phenomena (floods, ice bridges, gradual migration) or by human intervention. Although humans exist and operate within ecosystems, their cumulative effects are large enough to influence external factors—like climate.

What is a nutrient cycle?

Energy flows involve exchanges between an ecosystem and its wider environment. Nutrients primarily cycle within the ecosystem, creating exchanges among plants, animals, microbes, and the soil. Nitrogen, for example, is a crucial source of plant nutrition. It becomes available when soil bacteria convert atmospheric nitrogen to nitrate, but most land-based ecosystems have a limited capacity to capture nitrogen from the atmosphere. Hence, nitrogen cycling is an important determinant of ecosystem carrying capacity for plant species. Plant cells perform further biochemistry that renders nitrogen into a nutrient that can support other forms of life including insects and vertebrates. Plant tissues that are eaten or shed supply various forms of nitrogen to animals and microbes. Manure and the decomposition of dead organic matter return the nitrogen back to the soil. Although some nitrogen is returned to the atmosphere throughout these processes, this cycle can continue indefinitely so long as the nitrogen flowing out of the ecosystem is offset by nitrogen coming into the ecosystem through the work of nitrogen-fixing bacteria. Granted, this is a simplified picture of the nitrogen cycle, but other key nutrients follow cycles that are similar in form and function to that of nitrogen.

In ecosystems without human activity, the nutrient cycle will be spatially constrained by mechanisms for transport of nutrients from one organism to another. Winds and ocean currents provide one mechanism, while movement of animals is another. For example, flying animals (birds and bats) deposit large caches of nitrogen in the form of guano at specific locations, but terrestrial animals—toads, rabbits, or bears, for example—are far less mobile. Manure from these less mobile animals will become part of a more localized nutrient cycle. The pace at which nutrients cycle will be constrained by energy flows; ecosystems in tropical regions will have more solar energy available to power the metabolic processes that convert nutrients than those in cooler regions. Faster plant growth allows for more frequent turnover of nutrients, leading to an overall increase in carrying capacity. So the composition of organisms within a given ecosystem and the pace of nutrient cycling have significant controlling influence over the total number of organisms that can be supported at any given place over time. These factors determine the populations that can be sustained, while the continuity of energy flows and nutrient cycles determines whether the system as a whole is sustainable.

To get a sense of how nutrient cycles might be disrupted, consider a real-world case of human impact. As humans enter ecosystems, they manage nutrient cycles through agriculture. Many farming systems maintain the kind of geographically localized nutrient cycling just described, but as human populations began to concentrate in cities, crops and livestock were transported away from the farm, creating a gap in the nutrient cycle. Over centuries, farmers learned to transport nutrients over increasingly large distances. The discovery of island guano deposits in the nineteenth century was a crucial phase in the development of industrial agriculture. Transport of guano allowed farmers to fertilize crops (supply them with nitrogen and phosphorus), replenishing nutrients that were being removed from the farm ecosystem when crops and livestock were shipped long distances to urban centers. Massive deposits of guano were exhausted by the early twentieth century, but a method for fixing atmospheric nitrogen into synthetic fertilizer was invented by Fritz Haber (1868–1934) and perfected by Carl Bosch (1874–1940). Unlike natural nitrogen fixation by microbes, the Haber-Bosch process must be powered by an external source of energy.

The introduction of synthetic fertilizer was thus an important step that transformed agricultural ecosystems (i.e., farms) from the closed-loop nutrient cycling observed in natural ecosystems to one that requires continuous inputs from fossil fuels. Today, nutrients from crops are transported to other locations where they are consumed by livestock, which are then transported to slaughter facilities. The meat is transported yet again to urban consumers. It is not currently economically feasible to bring all wastes from the endpoint of this nutrient flow back to the fields on which crops are grown (although this does happen at a smaller scale in some communities where nutrients from treated wastewater are used as fertilizer). Thus, although the function of agriculture is to produce food, just as natural ecosystems produce food for the species that inhabit them, nutrient flows in contemporary food systems do not cycle. The closed-loop cycling that is crucial to the sustainability of a natural ecosystem is not replicated in the human food system.

What are stocks and flows?

Stock-and-flow relationships provide an abstract and generalizable way to understand many of the complex interactions that are sustained in ecosystems. We've just described energy flow in ecosystems: the movement of energy from the sun into forms that are used for the life processes of organisms. In describing the nutrient cycle, we've discussed the flow of nitrogen from the atmosphere into soil, to plants and animals, and back to the atmosphere or soil again. It is also important to notice junctures in an ecosystem where energy and nutrients are stored and accumulate. With respect to energy and nutrients, the storage points are the organisms themselves: plants, animals, and microorganisms such as bacteria. As the discussion already given shows, the leaves, roots, and stems of plants all store energy and contain nutrients that will eventually become available to the animals or microorganisms that eat or decompose them. Energy and nutrients will then be stored in the tissues of these organisms until they again become available for plant growth. *Stocks* are simply the storage points where flows accumulate for a time. In the examples we've been discussing, plants, animals, and microorganisms are stocks of energy and nutrients.

Identifying key stocks and flows provides a way to define an ecosystem. One can, for example, think of a population of organisms—a herd of buffalo or a stand of pine trees—as a stock that is replenished by flows of new organisms (buffalo calves and pine seedlings) and depleted when organisms die off. In this way of thinking, the birth rates and death rates control inflows and outflows. Inflows increase the stock while outflows reduce it. In fact, the very word stock derives from its use to describe populations of animals (e.g., livestock) that grow and shrink as individual animals are born and die (or are slaughtered, in the case of traditional livestock). In a similar vein, businesses think of their inventory as a stock. All of these various uses are consistent with the abstract notion of stocks and flows. In each case, the interaction of stocks and flows is a process that can be assessed in terms of sustainability.

The specific stocks and flows that one chooses to examine will constitute a particular *model* of an ecosystem. In the examples we've been discussing so far, we chose to concentrate on energy from the sun and on just one nutrient, nitrogen. Obviously, quite a bit more is going on in ecosystems than this, but by choosing just a few stocks and flows, we emphasize concepts of energy flows and nutrient cycling. In earlier examples, we chose to focus on deer and wolves. Again, obviously any real ecosystem has many more organisms than deer and wolves, but with a model in which there are just two stocks (deer and wolves) we are able to illustrate and study an important sustainability relationship between predator and prey that can be generalized to other stocks (other organisms) in an ecosystem. Modeling is one of the key activities in modern science, and it is not limited to stocks and flows. In general, a model is developed to simplify a picture or situation that would be overwhelmingly complicated and impossible to comprehend otherwise. The model emphasizes the important things and leaves other things out. Naturally, it is always worth asking whether you have left out something that is actually quite important. This is not a book that explains scientific methodology, but this concern points out how the real system that matters might differ from the way a scientist chooses to model it.

What is feedback?

Although there are many complex relationships in ecosystems, feedback is particularly crucial for sustainability. Feedback occurs whenever the size of a stock determines the size of a flow. This is most obvious in the case of population stocks like the ones just described above. The larger the size of a buffalo herd, the more calves will be born, and the more buffalo there are to die too. If the number of calves being born is roughly equal to the number of buffalo that die, the stock will remain stable. This kind of relationship between a stock and its inflows and outflows represents an *equilibrium* in which dynamism (or change) within the population (i.e., individuals dying and being born) nevertheless leads to a stock (a total population) that does not change. When the level of a stock influences inflow and outflow so that they produce an equilibrium, it is called a *balancing feedback*. An ecosystem in which inflows and outflows are balanced provides a paradigmatic example of sustainability.

In natural ecosystems, the balance between inflows and outflows may be regulated by complex relationships where the stock of one population (let's say wolves again) creates feedback to the outflow of another population (deer). As the stock of wolves increases, they kill more deer, increasing the outflow from the deer stock. As the wolf numbers decline, the outflow from deer declines as well. Thus predators (in this case wolves) are said to act as a control on the population of prey (deer). This is modeled as balancing feedback. But there are feedbacks that run between the stock of deer and the outflow for wolves too. As the number of deer grows, food for wolves increases, reducing outflow from the wolf stock or wolves dying from starvation. Yet even this is too simple, as deer are predators for the plant species on which they feed. Modeling the complex interaction of predator-prey relationships in ecosystems is one of the foundational tasks for ecological science.

But notice that if births are not offset by deaths, the size of the stock will grow, sending feedback to the inflow (births) making *it* grow, which only makes the stock grow faster and faster. The feedback between population size and birth rate is a *reinforcing feedback* that accelerates the pace.

Feedbacks occur in abiotic elements of an ecosystem too. This idea of reinforcing feedback is ecologically important in models of climate change. For example, scientists believe that as much as one-third of all soil-bound carbon is stored within the frozen tundra in the Arctic and Antarctic regions. With global temperatures on the rise, large areas of tundra are thawing and releasing carbon dioxide (an important greenhouse gas that helps warm the planet) into the atmosphere, reinforcing global warming. Also, as described in the next chapter, rising ocean temperatures reinforce rising air temperatures when heat energy is released by oceans into the atmosphere.

What is population ecology?

Thomas Malthus was writing about population ecology in 1798. He predicted that food production could not grow fast enough to sustain a growing human population. Malthus' point has been generalized to all sorts of resource consumption, and periods of resource scarcity have led environmentalists to predict dire consequences of an ever-growing human population. However, from an ecological perspective, humans are just one species among many.

In ecology, a population is a group of organisms that reproduce within a given ecosystem. Go back to what was just said about stocks, flows, and feedbacks for a herd of buffalo. Population ecology is a subfield that models these relationships in more detail. If you stick with a buffalo herd, notice that not all the animals are going to be contributing to the birth rate. The stock of female buffalo (cows) of calf-bearing age determines feedback for the birth rate, rather than the total population. Similarly, the stock of elderly buffalo determines feedback for the death rate. If the population of our buffalo herd has a disproportionately larger stock of calf-bearing cows as compared to elderly buffalo, the reinforcing feedback from calf-bearing cows will dominate the balancing feedback of elderly buffalo, leading to a growth in the total population of our buffalo herd.

One might think that the age distribution in a buffalo herd will be even, but there are natural events that can upset it. A late freeze, for example, can kill off a disproportionate number of young, or a virus may kill off a disproportionate number of old. These kinds of events are good examples of what ecologists call *perturbations*: one-time shocks that upset the balancing feedbacks that normally stabilize populations within an ecosystem. When this kind of disequilibrium occurs in nature, the population growth is temporary. Eventually, limitations in the food supply (buffalo prey on grasses) will limit the size of a stock, and balancing feedback dominates reinforcing feedback. Of course, buffalo and grasses are only two of the myriad number of species that can be found in any ecosystem. Population ecologists build complex stock and flow models to help them understand how the number of plants or animals in a prey species creates feedback that influences the number of animals in one of that species' predators, and vice versa.

What are predator-prey relationships in ecology?

In ordinary language, people tend to think that predators and prey are mutually exclusive: the predators out there are big and mean, while prey are meek and mild. *Tyrannosaurus rex* is sometimes described as "the ultimate predator." In ecology, a species can be both predator and prey. The term predator is applied generally to populations of organisms that feed on organisms of a different species. What does get carried over from ordinary use is that prey are the food of predators. If organisms in a population (e.g., buffalo, wolves, human beings) eat organisms from another population (e.g., grass, fish, buffalo), then those two populations exist in a predator-prey relationship. You might not think of a cow as a predatory species, but since cows eat grass and other plants, they exist in a predatory relationship to those plant species. Even large predators (like *T. rex*) are prey to tiny microorganisms. These relationships are important in ecology because they determine important feedbacks that make ecosystems sustainable.

How do predator-prey feedbacks shape ecosystems?

Buffalo eat grass, so grass species are prey for buffalo and buffalo are predators for grass. Wolves will hunt and eat a buffalo calf, so wolves are predators for buffalo. If you use the idea of a stock to understand the number of grass plants, buffalo, or wolves in a given region, you can begin to grasp ecological relationships in a simplified manner. As we've already said, lots of grass means the buffalo thrive, which we translate into stock and flow language as follows: the grass stock creates feedback to the buffalo outflow (i.e., their death rate). That is, when the grass stock is high, the buffalo death rate decreases; when the grass stock is low, the buffalo death rate increases. But there's also a feedback between wolves and buffalo: lots of wolves (a large stock) increases the death rate of buffalo; few wolves (a low stock) decreases the death rate of buffalo.

Notice also that the buffalo create feedback to the grass much as the wolves affect the buffalo. Lots of buffalo, and the grass death rate (from being eaten) goes up; fewer buffalo, and the outflow (or disappearance rate) for grass goes down. An ecological model will include the predator-prey feedback relationships of many species that exist in a nexus of these predator-prey feedback relationships. Buffalo eat a variety of plant species, for example, and wolves hunt and eat prairie dogs, deer, and rabbits as well as buffalo calves. Keeping all of these relationships straight is a complex task, but mathematical equations and computers allow ecologists to understand how a perturbation in one species (e.g., a flood or a drought) reverberates through many other species because of the feedback that is created by predator-prey relationships.

What is resilience?

We answered this question in chapter 1, and we discussed how business managers emphasize resilience by responding to shocks in chapter 2. But it is important to answer at more length in the context of ecology. Ecosystems are said to be resilient when they possess a nexus of stocks, flows, and feedbacks that return to an equilibrium after perturbation. The population of a herd may grow without limit after a disturbance that creates a disproportionate number of calf-bearing cows, and it may grow beyond the ecosystem's carrying capacity (i.e., beyond the availability of prey grasses to feed the herd). The herd population may then crash or decline below its typical numbers. These *oscillations* in the size of the stock may occur for a period of years, but if the feedbacks between stocks (buffalo and grasses) eventually reach a balancing point, the system will recover and demonstrate its resilience. In some cases, one species (one kind of stock) will emerge to assume the role that had previously been played by a different species. A species of clover may perform the function of providing food where previously that function was performed by a grass, for example. This illustrates the sense in which ecologists speak of an ecosystem recovering its functions or functional elements, as distinct from literally returning to its previous configuration of stocks, flows, and feedbacks. Ecosystems may thus demonstrate resilience while also exhibiting a fair amount of change in the specific plant and animal species that comprise them.

When ecologists talk about sustainability of ecosystems, they usually mean whether an ecosystem is resilient. They are measuring or estimating the sustainability of ecological processes in a given region in terms of whether a nexus of representative stocks, flows, and feedbacks will eventually return to something resembling their previous state after events that cause dramatic disruptions in the stocks (i.e., populations) of the organisms that define the ecosystem's structure. Some perturbations would push ecosystems beyond their ability to recover. Scientists believe that an asteroid crashed into the earth about sixty-six million years ago, causing perturbations in virtually all terrestrial ecosystems that led to the extinction of many species, including non-avian dinosaurs. In contrast, the 1883 eruption of Krakatoa created widespread perturbations resulting from tsunamis and dust that blocked sunlight, but many ecosystems were able to recover from this event. Ecologists' understanding of resilience and the factors that contribute to it are evolving. The rest of this chapter examines some ideas that are key for sustainability.

How does the sustainability of ecosystems become threatened?

The sustainability of ecosystems is threatened when they become less resilient. Identifying the factors that strengthen and weaken resilience has been a major theme in recent ecology. Not surprisingly, much of this work places primary emphasis on human activities that threaten resilience. Some ecologists have studied instances where specific ecosystems have gone into a state of collapse, leading to the disappearance of large human population centers and cultural systems. Others are working on more abstract criteria for system resilience. Yet another approach is to generalize longstanding work on carrying capacity and sustainable yields, discussed below.

How are ecological principles applied to identify threats to the global ecosystem?

This question is very much on the mind of contemporary environmental scientists, and there is no consensus on the answer. Ecologists have typically developed models of the stock and flow relationships in regional ecosystems—an island, lake, desert, prairie, or parts of oceans. While they recognize that the boundaries of these ecosystems are sometimes porous, the models do provide a measure of the extent to which predator-prey relationships can continue in a more or less stable fashion. That is, they can develop a measure of the sustainability of these ecosystems. But extrapolating to larger and more comprehensive ecosystems involves the hierarchy relationship discussed in chapter 2. It is analogous to the relationship between the sustainability of a single business and the sustainability of the economy as a whole.

Just as an economy can be sustainable when some of the firms within it are not, a more comprehensive ecosystem can be sustainable even if there are smaller subsystems within it that collapse. When the number of subsystems teetering on collapse reaches a tipping point, the more comprehensive system is also at risk of collapse. For example, a population may start to increase without limit when the predators that normally control the population go into sudden decline. This might trigger migration of that population in search of new food, and that process can, in turn, threaten other ecosystems. The reverse relationship in hierarchy can also occur. When functions within the more comprehensive system become unstable, the subsystems are subjected to more frequent and more severe perturbations. At the level of a large region, such as a continent, or the entire planet, energy flows can be altered due to fluctuations in the atmosphere following a very large volcanic eruption or collision with an asteroid. In both cases, the large number of particles in the air can block solar energy, leading to a lengthy period of cold. Some populations may not be able to adapt, leading to large-scale collapse in the predator-prey relationships that define more localized ecosystems.

One leading group has developed a theory based on *planetary boundaries*. The Resilience Institute is a global partnership of scientists focused on events that could disrupt ecosystems at a planetary scale and that, in their view, are under the control of human beings. Members of the Resilience Institute team, led by Johan Rockström and Will Steffen, identified nine areas where human activities could threaten the resilience of global ecosystems. They argue that ecological science should strive to estimate the threshold for change in each of these nine areas that would challenge the capacity for recovering ecological functions on a worldwide basis. The nine areas are: biodiversity, climate change, novel entities, ozone depletion, atmospheric aerosol loading, the global nutrient cycle, ocean acidification, freshwater consumption, and land use. If thresholds of activity or change in these systems can be estimated, Rockström, Steffen, and their colleagues argue that scientists can estimate the "safe operating space" for humanity. In other words, so long as these planetary boundaries are not exceeded, human life on earth is ecologically sustainable.

According to their current work, environmental scientists can make credible estimates of the safe operating space for humanity for seven dimensions, with the limits for aerosol loading and synthetic chemicals or engineered particles and life forms (i.e., novel entities) remaining largely unknown. Rockström and Steffen estimate that the limits for biodiversity loss and damage to the nutrient cycle have already been exceeded, with changes in land use and climate rapidly approaching the planetary boundary condition. Once boundaries are crossed, global ecosystems could become highly unstable. Interruptions in critical life-supporting stocks and flows could, in worst-case scenarios, range from economic shocks that would have life-threatening implications to climatic effects that would make regions of the planet uninhabitable. Nutrient cycles have already been discussed above. Further insight into the planetary boundaries approach can be gleaned by focusing on biodiversity.

What is biodiversity?

Biodiversity is one of the leading indicators that ecologists use to estimate the sustainability of both local and global ecosystems. It is a shorthand way of pointing to the fact that an estimated 8.7 million different species of plants and animals coexist on planet earth. Ecology itself points to the relationships among these species. Some species serve as food for others; some species play a role in other ecological processes such as soil formation or the nutrient cycle. Recall that most species eat several different kinds of prey species. When one species is affected, they can rely on another. But when a species disappears from an ecosystem (or when it goes extinct and disappears entirely), the predator's ability to adapt by switching from one food source to another permanently declines. This is, in a nutshell, why declines in the number of species in an ecosystem is said to reduce the resilience of the ecosystem. Beyond the nutshell explanation, it is important to remember that shocks can percolate throughout the ecosystem through the network of feedback relationships that exist among predators and prey. And on top of that, the loss of a species reduces the number of species that can perform tasks like soil formation or moving nutrients through an ecosystem. For some, the loss of species is particularly alarming because of how little ecologists really know about the role that most species play in global ecosystems and the associated ecosystem services. This means society has a limited understanding of what it could mean to lose them.

Biodiversity is declining at a historically unprecedented rate. Ecologists estimate that, under normal conditions, eight or nine of the earth's (estimated) 8.7 million species would go extinct each year, but human impact on ecosystem processes is now causing between one and ten thousand species extinctions on an annual basis. When all the different functions of species and their interactions are taken into account, a decline in the number of species (within an ecosystem or on earth generally) could be expected to affect the resilience of ecosystems at both the regional and planetary level. That's why Rockström and Steffen describe biodiversity as one of earth's planetary thresholds and why all ecologists regard the loss of species as an indicator of sustainability on a planetary scale.

How does climate change affect planetary boundaries?

The energy flows that shape climate at any given locale are one of the primary abiotic (i.e., nonliving) components of an ecosystem. Changes in these flows can have dramatic effects on the composition and balance of species. If average rainfall changes, it can increase or decrease the grasses and plants that are the food source for herbivores, for example. Unlike a volcanic eruption, these changes will be long-lasting, permanently modifying the composition of ecosystems and fundamentally altering their ability to support life. Changes in the stocks and flows of atmospheric gases that filter and regulate the flow of solar energy are predicted to have such extreme effects on ecosystems that many ecologists believe the resilience of every ecosystem on earth is now threatened by human influence. In short, the significance of climate change is so great that some environmental scientists now define the problem of sustainability in terms of mitigating the human activities that are forcing change in the stocks and flows that determine climate. As noted in chapter 1, we refer readers who wish to learn more about these challenges to Oxford University Press's book in this series *Climate Change: What Everyone Needs to Know®*.

What is applied ecology?

As you might expect, applied ecology is the use of ecology in making decisions that impact natural resources and wild populations of plants and animals. Applied ecology is important for thinking about sustainability because it is where the very idea of sustainability originated. The subfields in applied ecology go by many names that reflect specialized domains: forestry, game, and fisheries management; natural resource management; or simply ecosystem management. In fact, applied ecology is much older than ecology itself. Principles of applied ecology such as sustainable yield date to centuries before the German zoologist Ernst Haeckel (1834–1919) coined the word *ecology*, which he derived from the Greek word for "household." Although many people now think of ecology as predating sustainability, historian Ulrich Gruber argues the reverse. The German word for sustainability, *Nachhaltigkeit*, came into use in the early 1700s in association with procedures developed to manage the rate at which trees could be removed from forests in the Ore Mountains without damaging the forests' ability to repopulate. This was an early application of the applied ecology concept of "sustainable yield."

In short, non-scientists tend to think that scientists develop a theory (or model) and then apply it to a problem. In the case of ecology, however, dealing with problems that occurred in natural resource use came first. Sustainability came out of this tradition, and many of the concepts or tools that are now used in ecology predate the creation of the field as a scientific discipline. Sustainable yield is one of those tools.

What is sustainable yield?

The concept of sustainable yield was first developed in German forest management, and it has since been applied to fisheries and many other domains of applied ecology. Let's use the example of fisheries management as an illustration. The number of fish in a fishery (e.g., a pond, a lake, or an ocean) can be represented as a stock. It has an inflow that is determined by the natural rate of fish reproduction. The birth rate for fish will vary from species to species, but in the absence of hatcheries and other large-scale, human-managed systems for affecting stocks, it is a function of the size of the stock, which is, again, the number of fish that breed in an aquatic environment. As the stock grows, the inflow (or rate of fish coming into the stock) grows simply because there are more fish engaging in repro-duction. In the absence of some compensating increase in the outflow, the stock will grow faster and faster. The outflow (or disappearance rate) is the number of fish that are dying from natural causes or being removed from the stock by fishing spe-cies such as otters, bears, and human beings. Given what has already been said, the size of the pond, lake, or ocean and the availability of food for the fish will set a maximum size for the stock: too many fish and they start to die for lack of food and oxygen. It is also possible to take so many fish out through fishing that the replenishment rate (the inflow) becomes very small and the total stock will then take many reproductive cycles to build back to its maximum. That is overfishing.

Fisheries management evolved to control overfishing. The trick for fisheries management is to keep the stock high enough so that the inflow (fish birth rates) replenishes all the fish that fishers are taking out, as well as the natural death rate for fish (which may be eaten by other species in the pond, lake, or ocean). If one knows enough about the reproductive rate of a given fish species and other factors in the fishery's ecology, one can calculate this number and set a limit for each individual fisher that ensures the fishery as a whole will not decline. This principle has been known as sustainable yield in fisheries man-agement for almost a century, and if Gruber's history is right, a similar approach to calculating the rate for cutting timber in forestry has been in use for three hundred years.

Sustainable yield is therefore a foundational principle for understanding sustainability in the context of ecology. This is not quite where the explanation ends, however. Natural re-source managers have not only been interested in sustainable yields; they have tried to calculate the maximum sustainable yield. That is, they have tried to get as close to the tipping point as possible, taking the maximum number of fish, trees, or what have you out of an ecosystem without undercutting the population's ability to reproduce itself. This has proven to be a very tricky calculation to make, especially in large ecosystems with a complex mix of species and the potential for other disruptive events (like flooding, disease outbreaks, or drought). This attempt to hit the maximum falters because of the uncertainties involved in managing complex systems. It also falters because the managers are trying to manage human behavior, and humans are very good at finding ways to cut corners or cheat when other people try to manage their behavior. As a result, systems that have been managed according to a principle of maximum sustainable yield have experienced repeated failures. The collapse of the cod fishery in the North Atlantic is perhaps the most notorious failure, but it is only one of many.

Is sustainability a controversial idea in ecology?

Yes. One point of dispute concerns the definition and use of sustainable yield calculations. Sometimes applied ecologists have overestimated their ability to predict how the ecosystems they manage will behave, or the loggers, fishermen, and other users have not complied with their recommendations. Ocean fisheries have proven especially vulnerable to overuse and collapse of fish populations. Since the crash of the North Atlantic cod fishery off the coasts of Labrador and Newfoundland in 1992 and through subsequent population collapses in other parts of the fishery, the Canadian Department of Fisheries and Oceans has tried numerous ways to limit access and promote regeneration of cod stocks. However, populations remain low, and Canada further reduced fishing limits in 2018. Such management failures have led some ecologists to take a jaundiced view of sustainable yield as a concept in ecosystem management. In starting from the premise that managers are trying to sustain a maximum yield of harvested resources (e.g., timber, fish, game), the idea of sustainable yield places too much emphasis on resource extraction and not enough on resilience. In fact, the US Bureau of Land Management recommended abandoning some range management policies that emphasized sustainability and recommended a new approach based upon resilience.

The meaning or definition of sustainability within ecology is another point of controversy. At one time, ecologists thought that undisturbed ecosystems would go through a succession of stages, eventually reaching a stable equilibrium of species in balanced predator-prey relationships. Today, ecologists do not endorse the idea of climax ecology and the belief that ecosystems will be highly stable once they reach climax. The belief that an ecosystem, if left undisturbed, will continue to reproduce itself without substantial change in the composition of species is false. Ecosystems are in a constant state of evolution and change, partially in response to climatic change, but also simply due to the natural interplay among predators and their prey. If sustainability is thought to imply a stable, unchanging form for natural ecosystems, the idea derives no support from contemporary ecology.

Nevertheless, ecologists do rely on models of stock and flow relationships, feedbacks, and hierarchies to develop measures of ecosystem health. While episodes of collapse or rapid shifts in the populations that occupy a particular region are no longer thought to be unnatural, it is still meaningful to model how predator-prey relationships control stocks and flows within an ecosystem and to see how changes at one level in the hierarchy reverberate through other levels. It is also still meaningful to measure the extent to which the populations within an ecosystem can rebound after an external shock, such as a fire or flood. While some working ecologists would not use the word sustainability to describe these measurements, they nonetheless use the approach described in this chapter as a baseline for estimating how human activity is adversely affecting the processes that establish feedback among populations of non-human organisms within the global environment. We look more closely at environmental impact in the next chapter.

4

SUSTAINABILITY AND ENVIRONMENTAL QUALITY

What is environmental quality?

Succinctly, environmental quality encompasses the biophysical features or characteristics of the environment that affect the health or welfare of human beings and other organisms either positively or negatively. This is a very broad answer that could encompass many things. The presence of toxins or disease agents in an environment can impinge upon the health of organisms, while for humans, at least, aesthetic features of the environment may also be very important for welfare and a person's ability to live a meaningful life. Discussions of environmental quality often emphasize things that have a narrowly biological impact on humans and other organisms. For instance, a number of US state governments have Departments of Environmental Quality (or something similarly named) that focus primarily on factors affecting the biological survival of human beings and other plants and animals. We include concerns about recreation and landscape aesthetics and their importance to human well-being in our discussion of environmental quality.

Until recently, efforts to protect, restore, or improve environmental quality lacked the systems thinking necessary for sustainability. Richer ecological knowledge of systemic links among organisms and the abiotic elements of an ecosystem has deepened our understanding of the way that certain interventions to improve environmental quality can have very short-lived effects, while others can actually have adverse effects over the long run. A sobering example is the historical (and in some cases ongoing) introduction of exotic species to manage environmental concerns, with often devastating results. A species of toad brought from Central America to Australia in the 1930s to control a beetle damaging fields of sugar cane is now an ecological disaster. The cane toads, which have no predators, in turn prey on many native animals and insects, some of which are facing extinction. The toads are especially vilified by Australians because, with their toxic skin, they kill pets that catch and eat them. The methods of systems thinking discussed in chapter 3 help biologists and environmental managers appreciate the sense in which the sustainability of an ecosystem is important above and beyond (and sometimes even contrary to) the health and welfare of individual organisms living in it.

Nevertheless, some characteristics that impinge on the health or welfare of human beings and other organisms may have very little to do with whether an ecosystem is robust and resilient or whether it possesses adaptive capability. For example, cholera was an environmental health issue in the nineteenth century due to unsanitary systems for urban water supply, yet *Vibrio cholerae* bacteria are not inherently harmful to the functioning of ecosystems they inhabit. The *Anopheles* mosquito transmits a one-celled parasite that causes malaria, a serious health hazard in tropical climates. Eliminating *Anopheles* mosquitos from malarial regions would be a boon to human health, but would it harm tropical ecosystems? Most ecologists think not, because many other species of mosquitos would continue to play the crucial role that these insects play in food webs. It is arguable that well-targeted mosquito control neither promotes nor retards the sustainability of tropical ecosystems. People might still discuss the example of mosquito control under the heading of sustainability simply because sustainability has, for some people, become a high-level word or concept for collectively naming the things in our environment that matter to us.

Environmental quality is a comprehensive notion that includes too many factors to discuss here, especially given the way that people use the word *sustainability* to cover almost everything. We emphasize systems thinking for environmental quality, but the specific topics taken up in this chapter should be viewed as examples or components of the way that sustainability connects to environmental quality, rather than a comprehensive treatment.

How does environmental quality affect sustainability?

There is a simple answer to this question and a more complicated one that is the main focus of this chapter. The simple answer is that many people have started using the word *sustainability* where people would have once talked about environmental protection. That is, where people might have once said that a given practice hurts the environment, they started to say that same practice is unsustainable. In this view, sustainability encompasses protecting and even improving environmental quality. It might be a way to ask how economic activity affects our ability to do this, but everything that would have been included in environmental protection is automatically transferred over to sustainability without modification or dilution.

This change in vocabulary is significant. By the 1990s, asking people to change their lives in order to protect the environment had become politicized. Environmental regulations were supported by people who tended to vote one way, while reducing regulation and government interference was championed by people who voted differently. The shift to talking about environmental quality in terms of sustainability gave the folks on the anti-regulation side of that conversation a new language for talking about changes in environmental policy and practice that they would support. Some environmentalists were willing to make this change of vocabulary too. There was a bit of easing in the political deadlock, and sustainability became a new way for people with otherwise antagonistic political views to discuss practices and policies for protecting the environment.

The shift in vocabulary may have enabled some more profound shifts in thinking about the environment, and that is the main topic of this chapter. Before moving on to that, let's note some less helpful dimensions of this change in the way people talk about the environment. First, it helps us understand why many people just assume that sustainability is concerned with the environment and nothing else. We countered that thought in chapter 1, but we recognize that for many people sustainability continues to be a way to talk about protecting the environment. Second, if a change of words is all that is going on, people on every side of the political issue would be justified in being skeptical about this new buzzword. It would just be a way of concealing important differences of opinion. "Eschew obfuscation" was a self-parodying bumper sticker back in the day. Anyone who supports that sentiment would have been leery of the new talk about sustainability.

We can start to consider the more complicated answer to our question by taking up where chapter 3 left off. There we saw how ecologists build models for understanding the relationships among living organisms and energy flows within a region. They think of populations and abiotic (i.e., nonliving) elements in terms of stocks and flows, and their models reveal how the levels of one stock trigger feedback to the flows affecting other stocks. Models of sustainable yield were used in managing fish and game for decades. As these models started to incorporate more stocks, flows, and feedback, they led to a new way of thinking about protecting environmental quality. This was a gradual change for professionals working in fields such as fish and wildlife management or environmental protection, so gradual that it may not have been obvious to them what was happening. For these insiders and specialists, the question is less "How does environmental quality affect sustainability?" and more "How does sustainability affect our approach to protecting environmental quality?" With that, we move into the heart of this chapter.

How is environmental quality different from ecology?

If you are reading this book from front to back, you've already read about how ecologists use the principles of sustainability to understand important natural processes. As we noted above, ecologists build models to explain the interconnection of stocks and flows in an ecosystem. But when people talk about environmental sustainability, they may not be thinking about natural ecosystems. They may be thinking about the way that pollution can threaten human health or the way that industrial processes have negative effects on the quality of life. They may be thinking about restoring places that have been degraded by dumping garbage or industrial waste or preserving parklands and habitat for native plants and animals. They may be thinking about recycling or reducing waste to minimize environmental impact. Importantly, they may be focused much more on specific aspects of environmental quality than on the ways in which plants, animals, and abiotic elements in an environment are systemically organized and the factors that make these ecosystems robust, resilient, and adaptive. In a nutshell, ecology aims to understand the causal connections that exist in ecosystems, while environmental quality is evaluated in terms of how well an ecosystem is performing for certain purposes.

In many cases, our standard indicators of environmental quality are systemically connected to the stocks and flows that make ecosystems sustainable, especially when one considers the planetary-scale ecosystems that everyone depends upon. But these interconnections can be very complex, and they can also be controversial. Therefore, in environmental policy and management it has often proven more effective to focus on impacts that are easy to see and on outcomes that people agree about: reducing diseases, such as asthma or cancer, that can be linked to air pollution; cleaning up waterways or restoring habitat for fish and game; or reducing the amount of waste that gets deposited in landfills. And for many people, these specific outcomes basically define what sustainability means.

In this chapter, we focus on some of the indicators that are widely associated with sustainability, but we don't look too closely at whether they actually contribute to our planetary or local ecosystems' ability to reproduce themselves over time. The goals that these indicators target probably are important to people because they improve their quality of life. Many of them were important components of environmental policy and activism well before the idea of sustainability appeared on the scene. There are situations in which it is important to ask whether environmental policy makes ecosystems more sustainable or not. For many people, seeking and achieving quality-of-life goals through environmental protection is very much at the heart of sustainability. However, no one should overlook the way in which the ability of ecosystems to provide critical services is very much at the heart of quality of life.

What are ecosystem services?

In chapter 2, we noted that ecosystem services are the ways in which humans rely upon and benefit from the things that natural systems provide. The concept of ecosystem services surfaced in the 1960s, but it was not until late in the 1990s that it received widespread attention following the publication of Gretchen Daily's book *Nature's Services* and a series of articles published by ecological economists that focused attention on the way in which human societies rely upon ecosystem services. When the UN secretary general, in 2000, called for an assessment of the consequences of ecosystem change for human well-being, attention to ecosystem services sharpened. Almost 1,400 social and natural scientists from around the world participated in the Millennium Ecosystem Assessment, which appraised the condition of the world's ecosystems and the services they provide.

The Millennium Ecosystem Assessment classified eco-system services according to four types of services: provisioning, regulating, cultural, and supporting. Provisioning services refers to the ways in which ecosystems directly provide products and services that humans rely on, including, among others, food and fiber, fuel, freshwater, biochemicals, and genetic resources. Regulating services include benefits like climate regulation, water purification, regulation of hydrologic systems, pollination, and biological control of pests and diseases. The cultural services category contains nonmaterial, non-extractive benefits that people realize from ecosystems—such things as knowledge systems, spiritual and religious values, aesthetic enjoyment, sense of place, and cultural heritage. Finally, supporting services refers to the ways in which ecosystem functions support the other three kinds of ecosystem services. Supporting services benefit humans indirectly; for example, water and nutrient cycling support the provisioning and regulating services provided by water and soil nutrients. This chapter focuses on the ways that pollution and resource depletion affect environmental quality—or what matters about the environment. But many of those things exist as ecosystem services. Even when the focus is on environmental quality, one is never far from the concept of ecological sustainability.

What is pollution?

Stated very generally, pollution is the contamination of any substance or process by some other substance or process. The opposite of pollution in this sense is purity or cleanliness. Although the idea of pollution has applications in the purifying rituals of religious practice, in the context of environmental quality pollution occurs when some environmental amenity such as water, air, or soil is contaminated by a toxin or other health-affecting substance or by something that reduces the ability to use or enjoy that amenity. Stated another way, pollution damages an ecosystem's capacity to provide critical goods and services that humans (and other plants and animals) rely upon. Health and enjoyment are values that are, to some degree, in the eye of the beholder. Hence, there is controversy over whether something is a pollutant or not. At the same time, public health science has established very robust accounts of causative agents in diseases including cancer, emphysema, and heart disease. Ecologists have identified the impact of toxic chemicals on the reproductive rate of fish and other species. In this chapter, we emphasize pollutants shown to have negative impacts on the health and reproduction of human beings and other species living in a given environment.

Three main ways pollutants cause damage are acute toxicity, chronic health risk, and endocrine disruption. Acutely toxic pollutants include poisons and other chemicals, as well as microorganisms, that injure or kill individuals within a matter of weeks, days, or even hours. These would include industrial chemicals such as chlorine or hydrogen sulfide or microorganisms such as salmonella or E. coli O157:H7. The toxicity of acute toxins varies by exposure or dose. Many acute toxins are ubiquitous in the environment at doses that are thought not to cause harm. Acute toxins pose serious hazards when exposure levels reach harmful thresholds, and environmental regulatory agencies are quite vigorous in responding to them. Nevertheless, it is precisely because chronic toxicity is more difficult to identify and longer acting that pollutants known or believed to have chronic health impacts are often more significant for environmental quality.

Endocrine disruption is a relatively recently discovered form of toxicity. Endocrine disrupters have a microscopic shape that mimics that of hormones or other substances produced naturally in the bodies of humans and animals. These natural substances may regulate the onset, degree, and termination of various natural processes, such as puberty or the ability to reproduce. Scientists know that endocrine-disrupting chemicals in the environment affect the development of fish and amphibians, but their impact on human beings is still debated. The science documenting the impact of endocrine disruption was hotly debated through the 1990s. Although endocrine disruptors are now recognized as potential pollutants, the science for measuring their impact on environmental quality is still being developed.

Littering and other human activities can also cause direct harm. Plastic bags, packaging, and netting entangle wildlife. Plastic straws clog nasal passages or cause internal blockages in animals that encounter them. Discarded trash can make a specific location uninhabitable for the creatures that lived there before it was polluted. The accumulation of plastics in the stomach and intestines of fish, birds, and other animals can be fatal, even though the materials themselves do not have toxic properties. Abandoned buildings or other human-built structures can disrupt the activities of animals living in nature. These examples illustrate the variety of ways in which humans can damage environmental quality for nonhumans living there.

Some environmental groups have advocated that greenhouse gases should also be viewed as environmental pollutants. These emissions have large-scale impacts on the amount and composition of sunlight, the hydrologic cycle, and other processes that regulate ocean temperature and storm cycles. Environmental agencies including the US Environmental Protection Agency (EPA) have, at various times, proposed regulating emissions under laws that govern pollution. Unlike acute toxins, chronic agents, endocrine disrupters, or other types of pollution, the greenhouse gas emissions driving climate change may not directly affect the health of individual organisms. Yet the cumulative effect of these emissions could make entire regions unlivable. The debate illustrates one important way in which contradictory value judgments influence our understanding of environmental quality.

What is resource depletion?

Resource depletion occurs when an ecosystem's ability to supply a good cannot keep pace with the rate at which it is being used. In the most obvious cases only a finite amount of the good is available, and every use reduces the amount that will be available for future use. Fossil fuels such as coal and petroleum are finite resources that are inexorably depleted by mining and use. When resources are potentially renewable but patterns of use damage the feedbacks that would support restoration of stocks, more subtle forms of depletion occur. The depletion of earth's renewable resources is a substantial component of environmental quality, because people who depend on the reliable supply of a resource will suffer once it is gone. The collapse of the Atlantic cod fishery (discussed in chapter 3) is a profound example of resource depletion in this sense.

Depletion is much more difficult to manage and regulate than pollution. The Aral Sea was once the largest lake in the world. Rivers that fed the lake began to be diverted for irrigation in the 1960s. By the 1990s, only 10 percent of the Aral Sea's surface area remained. The region once occupied by the lake continues to experience oscillations in the quantity, quality, and location of water. The depletion of this resource destroyed the livelihoods of fishing communities that had populated the region, though fish have returned to some of the remaining wet areas. Depletion of the lake adversely affected the quality of drinking water in the region, causing further impacts on public health. Although planners expected that the surface of the lake would decline when irrigation began, they did not anticipate many of the other problems. Specialists consider the depletion of water in the Aral Sea to be one of the twentieth century's greatest environmental disasters.

The capability of soils to support food production is a potentially renewable resource that could, in theory, allow for continuous production of crops. Crop residues and animal manures can replace nutrients taken from the soil in the process of plant growth. However, many modern farming practices substitute synthetically produced fertilizers or disrupt the processes that regenerate soil fertility entirely. Some forms of cultivation allow significant erosion of soil by wind and rain, which reduces soil productivity and deposits soil nutrients and sediment elsewhere. Soil nutrients and sediments washing into streams become pollutants that contaminate water, threaten aquatic ecosystems, and interfere with human activities like drinking water treatment and navigation. A more subtle form of depletion occurs when grains are shipped to animal production facilities, and then animal products are shipped to urban centers around the globe. These shipments are effectively transporting nutrients that could otherwise be replenishing soils (see chapter 3 for a discussion of the nutrient cycle). While synthetic fertilizers have thus far proven capable of maintaining crop yields, changes in soil characteristics have led advocates of sustainable agriculture to question whether this lack of integrity in nutrient cycling can continue indefinitely.

How can environmental quality be protected?

Environmental protection is a large topic, so large that *Environmental Protection: What Everyone Needs to Know®*, by Pamela Hill, is available in the same book series as this one. Here, we focus on the two aspects of environmental quality we've discussed already, pollution and resource depletion. The obvious thing to do is pollute less and use fewer resources. One might think of these actions in terms of things people do at the household level: someone can shift from consuming products that pollute to products that don't (or that pollute less). The chemicals in laundry detergent are released into the environment every time someone washes a load of clothes. They might choose a brand of detergent that has less toxic and harmful chemicals. At the household level, one can reduce the load on resources by using less: turning down the thermostat to conserve energy or taking shorter showers. We discuss some of these strategies in chapter 9.

Businesses can do similar things to reduce pollution and resource depletion, and we discuss some of those things in the rest of this chapter. Local, regional, and national governments can develop policies that protect the environment by encouraging conservation and by banning the use of particularly harmful substances altogether. However, all these activities involve cost. At the household level, accounting for the costs and benefits of environmental protection matters for your household budget, but when one moves from the household level to that of a business or a government, cost-benefit accounting becomes critical, because wasting money on ineffective measures violates the trust that people have placed in those who make decisions intended to protect environmental quality.

For our purposes here, we skip the economic theory behind cost-benefit accounting for environmental protection, but we can't skip over the need for observable quantities that indicate whether measures being taken to protect environmental quality are working. With pollution and resource depletion as our examples, the obvious place to start is the amount of polluting materials being released and the amount of resources being depleted. Here, pollution is treated as a stock that is going up as a result of human activity, while resources are treated as a stock that is going down. To use the concepts we introduced in chapter 3, the stock of pollution is rising because human-generated wastes function as an inflow, while resources are being depleted because human use is an outflow. This is only a starting point, and it's a drastic oversimplification. Some pollutants are more toxic than others, and as discussed earlier, pollutants affect environmental quality in many ways. The quantities scientists measure to assess impacts on environmental quality need to reflect these differences. The quantities used in environmental science are referred to as *indicators*. And good indicators help us decide where to intervene in systems (systems thinkers call these points of intervention *leverage points*) to reduce unwanted environmental damages and make environmental improvement efforts more successful.

What are environmental indicators?

An environmental indicator is a measurable quantity that tells us something about environmental quality. Some indicators assess the actual state of the physical environment, for example, the concentration of a pollutant in a lake or stream. Such indicators are called state indicators. Environmental and public health specialists care about phosphorus concentrations because high levels of phosphorus in lakes and streams cause algal blooms that may be harmful to human health and to fish and other aquatic organisms. Other types of indicators assess the actual effects on humans or ecological systems of declines or improvements in environmental quality. These are called impact indicators. For example, many areas monitor the number of individuals reporting to emergency rooms in respiratory distress as part of air quality monitoring programs; research in a number of cities has shown that the number of asthma cases treated at local hospital emergency rooms increases when concentrations of particulate matter in the air are high.

A given quantity may become an indicator of environmental quality for different reasons. Some indicators are selected because they are informative; an informative indicator helps people understand environmental quality and enables them to think about how their own behavior might affect it. Informative indicators are linked very closely to the condition of interest. An indicator may also be selected because it is reliable: even if it is difficult to see how the indicator is connected to broad notions of environmental quality, research has shown that it is closely correlated with some environmental quality goal. As such, it is a good tool for hands-on decision-making because it helps managers and policymakers accurately predict the impact of a given decision. The air quality–human health example is a good one here. On the one hand, measuring the actual concentration of particulate matter in the air offers an informative indicator; scientists and decision makers can observe how certain industrial practices, vehicle emissions, wildfire, and other activities known to emit particulates into the air affect air quality. On the other hand, number of reported cases of asthma at local hospitals is a fairly reliable indicator of air quality.

Ideally, an indicator is both informative and reliable. An indicator that is neither informative nor reliable is worse than useless because it can cause us to do the wrong thing. To the extent that improving or maintaining environmental quality is part of sustainability, it is important to have a mix of both informative and reliable indicators. While reliable indicators may be the most useful tools for day-to-day decision-making, people need informative indicators that provide an understanding of how management decisions are directly affecting environmental quality. Informative indicators provide a clear picture of environmental quality and may result in broader endorsement of environmental quality goals by people whose loyalties span the entire range of the political spectrum.

Lest we lose sight of why environmental indicators are important tools in discussions of sustainability, remember that most indicators are measuring quantities. State indicators are measuring the size of stocks and flows. Impact indicators are measuring how the size of one stock affects flows into or out of some other stock (i.e., feedback). If concerns about environmental quality reflect concerns about whether pollution or resource depletion somehow limit important processes or practices (the interactions of these stocks, flows, and feedback), then scientists must be able to assess how everyone's actions affect these processes and how changes in these stocks and flows affect society and public health. When one moves from protecting environmental quality simply by trying to reduce the flows that contribute to pollution or resource depletion toward examining how many factors are connected through feedback, one is starting to view environmental quality from the perspective of sustainability.

Some common environmental indicators measure freshwater quality, outdoor air quality, climate change, ocean temperature, land use change, and waste generation. Importantly, these indicators matter not just because they tell us about water quality or land use, for example. Because of feedback relationships, indicators used for each of these also provide information about a host of other variables that people care about.

What do indicators of freshwater quality tell us?

The example above—the concentration of phosphorus in lakes or streams—is an example of a water quality indicator; specifically, it is used to assess the ability of fresh surface water bodies to support aquatic organisms. A host of chemical and biological contaminants have found their way into surface water bodies, and environmental protection agencies monitor the levels of those contaminants that are known threats to human or ecological health.

Many other freshwater quality indicators do not involve measuring the concentration of specific contaminants in water bodies at all. Instead, for example, scientists and regulatory agencies monitor aquatic species that are found, or should be found, in freshwater to assess water quality. Fish kills, observed abnormalities in fish, or the presence of contaminants in the flesh of fish species are biological indicators that scientists use to assess water quality. They also assess water quality based on the presence or absence of aquatic macroinvertebrates, such as insects, snails, or worms. The number of different species present, the specific species observed, and the number of individual organisms counted provide information about how water quality is or is not supporting aquatic life. Scientists can use the amount and types of aquatic vegetation as an indicator of water quality as well. Finding vegetation that is undesirable or failing to find beneficial aquatic plants provides information about what may be in the water that is aiding or preventing plant growth.

Water quality indicators for fresh water and salt water (oceans) differ, since biological and chemical contaminants respond differently in fresh and salt water. For example, while high phosphorus levels in fresh water result in excessive algal blooms (in the United States, Lake Erie water quality is a good example of what happens when phosphorus levels are too high), high nitrogen levels in saline and brackish water cause algae growth. The lesson here is that water chemistry matters. Similarly, whether water bodies primarily are used as drinking water sources, are recreational resources, or are valued for fisheries affects how water quality is measured. High phosphorus levels in surface water are not considered a human health risk, but they are a risk to fish populations. On the other hand, certain forms of nitrogen are a human health risk if they are found in drinking water. (In most countries with rigorous drinking water quality protection programs, public water suppliers must test for nitrate levels. However, rural residents with private wells should be cognizant of potential risks to their well-water quality, since they are responsible for making sure their drinking water is safe. High nitrate levels in private wells, especially in rural areas, are disturbingly common.)

The percentage of the population whose homes are connected to public wastewater treatment facilities is an indicator of water quality that doesn't measure anything that is actually in the water. This number provides information about the degree of risk to freshwater quality from human waste generation. The international Organisation for Economic Cooperation and Development (OECD) publishes data collected for many environmental indicators, including this indicator of freshwater quality. In the early 1980s, about 50 percent of the population in OECD member countries were connected to a municipal wastewater treatment plant; that number rose to almost 80 percent by 2015. Whether this particular indicator is informative or reliable requires us to dig a little deeper into what is happening in the communities where the data is collected. Some locations have more advanced wastewater treatment technologies in place, so risks to freshwater quality are fewer. On the other hand, in some places wastewater infrastructure is old and badly in need of improvements, so water quality is suffering. We look more closely at how governmental efforts to protect environmental quality are addressing sustainability goals in chapter 7.

How is outdoor air quality assessed?

The levels of six common air pollutants are monitored across the United States, and standards have been established by the EPA to protect the environment from damage by high concentrations of one or more of these contaminants. The six criteria pollutants are carbon monoxide, lead, nitrogen dioxide, ozone, particulate matter, and sulfur dioxide.[1] Most of these pollutants are released directly into the air from sources like manufacturing, energy generation through burning coal and natural gas, and automobile engines, among others. Ground-level ozone is formed when various chemicals released into the atmosphere react in the presence of sunlight. The OECD air quality monitoring program uses emissions of sulfur oxides, nitrogen oxides, and particulates as indicators of air quality, rather than measures of actual concentrations of these contaminants in the air. For member countries, OECD calculates the levels of these emissions based on the types and amounts of economic activity and the use of various energy sources in each country, as well as the extent of pollution abatement technology adoption (such as catalytic converters on cars and control technologies used by various types of industries). Around the world, air pollution control programs use urban traffic density, the number of days air quality fails to meet regulatory standards in a given city or region, and the adoption and enforcement of regulatory programs for protecting air quality by state and national governments as air quality indicators.

Outdoor air quality is a concern because public health, natural systems, and the built environment can be negatively affected if certain airborne contaminants exist in high concentrations. We've already mentioned the connection between high levels of particulate matter and respiratory ailments. Carbon monoxide and ozone in the lower atmosphere also result in respiratory distress. Many cities report air quality alerts on days when levels of these pollutants are expected to be particularly high and suggest that residents limit outdoor activities. Large areas of forestland across the globe have been damaged by the deposition of airborne contaminants through acid rain. In many cities around the world, buildings and statuary have been damaged by acid rain. All of these are examples of how poor air quality threatens the sustainability of important practices and processes. They are also reminders of why systems thinking is important. Discussions of healthcare costs rarely include discussions of air pollution, nor do expressions of concern about depletion of forest resources.

What do indicators tell us about climate change?

Climate change indicators include both state and impact indicators. That is, there are indicators that help us understand the current state of key climate variables, and there are indicators of the impacts that climate change is having on the environment and on human health. The principal driver of climate change is the increasing concentration of certain gases—called greenhouse gases—in the earth's surface atmosphere. The presence of these gases in the atmosphere causes the greenhouse effect—a natural process in which the gases trap some of the sun's energy that is radiated by the planet, causing the atmosphere near the earth to warm and making the planet habitable. The increasing concentration of greenhouse gases in earth's atmosphere has meant more heat trapped near the earth's surface and higher air temperatures. For this reason, the term *global warming* has often been used interchangeably with the term *climate change*. However, global warming refers specifically to an increase in air temperature near the earth's surface, and this is but one manifestation of a changing climate.

One major indicator of climate change is the direct measurement of the concentration of greenhouse gases in the earth's atmosphere. These measurements are taken at monitoring stations around the world. Greenhouse gases are also monitored by tracking the emission of greenhouse gases, which is measured directly or calculated based on the amount of fossil fuels burned or other activities that emit greenhouse gases.

Those who study climate change also use a series of indicators specifically related to weather and climate. Meteorologists distinguish between weather and climate like this: weather is what is going on at any given time and place (including things like temperature, precipitation, clouds, and wind), and climate is the long-term average of weather in a given place. However, climate is not just defined by long-term averages of things like temperature and rainfall. Climate also includes things like the frequency, duration, and intensity of heat waves, cold spells, storms, and droughts. Some of the common weather and climate indicators include long-term temperature averages as well as trends in high and low temperatures, trends in average precipitation, frequency and severity of storm events, size and frequency of river flooding and flooding in coastal regions, and frequency and severity of drought. Measurements of the extent and melting of sea ice in the Arctic and Southern Oceans and of glaciers in polar regions and mountain ranges are also indicators of climate change.

Health-related indicators are also used to monitor the impact of climate and weather on human health. The number of heat-related illnesses and deaths, the rate of diseases like Lyme disease and West Nile virus, and the length of pollen season are examples. Tracking the extent of wildfires, changes in the winter ranges of birds, and the distribution of plant and animal species also provide data for climate change indicators. Among many other climate change indicators, ocean temperature is an important one—important enough to get its own section here. But one important concern about rising ocean temperatures is that the heat energy absorbed by oceans has to go somewhere, and much of it is released back into the atmosphere. This feedback, flows of heat from warming oceans back into the atmosphere, could add to the warming of the planet for decades after the heat was initially absorbed by the oceans. Remember how the concept of sustainable development (chapter 1) asks us to take the needs of future generations into account. This is but one example of the long-term impacts of decisions that people are making now.

Why is ocean temperature monitored?

It is hard not to envision people dropping little thermometers into the ocean all over the world when discussing this environmental indicator. In fact, scientists all over the world are measuring sea surface temperature (SST), although what qualifies as the surface varies from a few millimeters deep to as much as two kilometers deep. They obtain some measurements from satellites that capture temperature data and other information from thermometers mounted on networks of floating buoys. Ships measure SST and compile the data for national agencies that collect and analyze the information. Marine telemetry is also used to obtain temperature measurements: sensors attached to marine species of all sizes transmit data on temperature and other ocean environment variables, as well as information about the animals' interactions with their environment. Satellites mounted with sophisticated infrared imaging technology measure SST directly.

That's how ocean temperatures are monitored, but what kind of an environmental indicator is ocean temperature? What does ocean temperature tell us about our environment? Marine ecosystem health is temperature dependent, so SST is an important piece of information for commercial fisheries management (one of those variables that affect decisions about sustainable yield, discussed in chapter 3) as well as for tracking species like sea turtles and monitoring the health of coral reefs. Warmer ocean water means more storm systems, so monitoring ocean temperatures helps with predicting rainfall. Monitoring ocean temperatures also helps anticipate El Niño and La Niña cycles, which affect weather patterns. The tourism industry watches ocean temperatures closely. And gradual warming of the ocean's surface reflects the overall warming trend within the global climate system. Monitoring SST helps us understand how oceans are responding to climate change and how that response may in turn affect a host of environmental factors (aka ecosystem services) that are important to all of us.

What does monitoring land use change tell scientists about the environment?

The uses made of land resources can significantly affect the environment. Environmental scientists know that large tracts of forest land and other plants serve an important role in carbon sequestration when they remove carbon dioxide, a principal greenhouse gas, from the atmosphere during photosynthesis. They also know that flood risks increase as the amount of land area covered with impervious surfaces, including roads, parking lots, driveways, and rooftops, increases. Smaller patch sizes for wildlife habitat as well as a loss of connectivity among patches threaten species dependent upon that habitat. Parcelization of privately owned agricultural and forest lands threatens their ability to produce the goods and services people depend upon.

Researchers use several databases to track land cover, which is simply the physical properties of the land surface. Land cover includes things like grass, trees, agricultural crops, impervious surfaces, wetlands, and open water. In contrast, land use databases track how people are using the landscape, such as for agricultural, residential, commercial, recreational, or other uses. Clearly, land use and land cover are closely related. Satellite imagery, aerial photography, and ground-level surveys collectively provide information on land cover and land use. Extent and type of land cover and land use are indicators that help resource managers predict impacts of floods and storm surges, assess potential impacts from rising sea levels, assess and respond to wildfire risks, and make decisions about wildlife management. So monitoring land use change helps us monitor environmental quality, understand how our actions affect a host of interconnected systems, and make important decisions about environmental protection.

What do indicators of waste generation reveal about sustainability?

Two principal concerns drive the monitoring of waste generation. First, the amount of waste generated has important implications for how waste is managed. At the simplest level, where waste materials—household waste, industrial waste, or other kinds of byproducts of human activities—end up matters to people. Think about conflicts over the siting of sanitary landfills, air quality concerns related to waste incinerators, questions about the safety of hazardous waste disposal facilities, or even the decades-long debate about storage of wastes from nuclear power plants. Of course, everyone uses products that result in waste, whether it is the packaging they put into the garbage in their own homes or the waste created during the manufacturing of those products. But few of us are enamored of having a landfill or incinerator or hazardous waste storage anywhere near where we live, work, or go to school. In fact, the acronym NIMBY (not in my backyard) surfaced during the 1980s as a result of residents' efforts to prevent the siting of landfills and hazardous waste storage facilities in or near their communities.

The second concern arises from something that ecological economists call throughput. The term throughput is used in many fields, including business and data management, but its meaning differs slightly in each field. Ecological economists use the term throughput to refer to the flow of raw materials and energy from natural systems through the economy and back to the natural environment. Things like minerals, food and fiber, and even water are removed from the physical environment and used in other places, often after being transformed into something that people want. But a basic law of physics tells us that matter cannot be created or destroyed. So any physical materials that go into producing things people want have to end up somewhere, and eventually most of what anyone buys is going to end up as waste (even the antique dining room set or collection of tools that have been passed through the family for generations). However, throughput is important not just because of waste generated on the back end of economic activity but also because of the use of raw materials on the front end. Reliance on natural resources, especially nonrenewable ones, to produce goods people want means humanity is continually reducing the resource stocks that people rely upon. If people use renewable resources faster than they are regenerated, then these potentially infinite assets also become nonrenewable. Absent substitutes for natural resources everyone relies upon, the rate at which people consume resources matters.

In both of these cases, waste generation is an important environmental indicator. Even with great strides in recycling and emphasis on reuse, the availability of space in existing waste treatment and storage facilities is on the decline. Knowledge about rates of waste generation helps those in the waste management industry know how much longer society can rely on existing facilities. Finding sites for new facilities is difficult; NIMBYism is still alive and well. Data on waste generation, when combined with data on waste transport and the location and characteristics of storage facilities, helps with tracking the potential for negative environmental or human health effects from waste management (or mismanagement). Activists concerned about environmental justice (which we discuss in chapter 6) pay special attention to rates of waste generation, types of wastes, and transport and storage of wastes, because environmental and health risks affect people at lower socioeconomic levels disproportionately.

Waste generation indicators are also indicators of throughput. Data on the amount and types of wastes generated and where that waste goes represent one more way to track reliance upon and extraction of raw materials. Recycling and reuse are encouraged because they may reduce throughput. Intuitively, one would expect that recycling things like paper, aluminum, and plastic reduces pressures on renewable and nonrenewable resources because less of the raw materials (trees, aluminum, and petroleum) is needed. An important caveat is necessary here, though: absent a full life-cycle assessment, one cannot be sure that recycling always results in a net reduction in the use of raw materials. The processes for making some materials, such as cardboard, useable again require other inputs (e.g., water, chemicals) that generate other kinds of waste. Lessons from waste management have demonstrated that life-cycle assessment is a valuable tool in sustainability work.

What is life-cycle assessment?

A life-cycle assessment is an analysis of how a given product or activity affects environmental quality throughout all stages of its useful life, starting with the impact on natural resources from obtaining raw materials through its manufacture and distribution and moving on through its use or deployment and ending with the environmental impact of retiring the product and treating its material components as part of the waste stream. While most of the environmental indicators discussed above focus on relatively large-scale ecosystems, life-cycle assessment has become one of the most powerful tools for sustainability managers because it helps them think more comprehensively and systemically about the operations or activities of the firms or organizations they work for.

A life-cycle assessment will follow the product that a firm produces through each phase of its production, distribution, use, and retirement. At the production stage, the assessment will measure the resources being used to make the product, including both materials that go into it and the water and energy use and wastes generated during manufacture. Distribution measures similar impacts on the processes of transport, marketing, and delivery to users. The use phase also impacts environmental quality. An automobile, for example, will burn fuel, require tires and maintenance, and emit gases throughout its use phase. Finally, when the use phase is complete, a product will either be recycled or go into the waste stream, having further effects on environmental quality. A well-constructed life-cycle analysis can help managers identify the most effective leverage points for conserving resources and eliminating pollution. In many cases, not only have life-cycle analyses allowed firms to reduce harmful impacts on environmental quality, but they have improved the profitability (i.e., sustainability) of the firm itself. Comparative life-cycle assessments enable a firm to choose from among products or production processes to meet environmental quality goals.

For consumers, comparative life-cycle assessments can inform purchase decisions when choosing among alternative products. Let's consider, for example, the decision to purchase a pair of shoes. A comparative life-cycle assessment would consider where and how the materials used to make the shoes are obtained and their environmental impact. The decision to buy a leather shoe would involve determining what type of animal hide was used, how the animal was raised, and the use and disposal of chemicals involved in tanning and processing the hide. Alternatively, the decision to buy a shoe made from a synthetic faux-leather material would involve understanding how that material is produced; many such materials are made using petroleum-based products, which requires understanding the extraction and refining of petroleum in addition to how the materials are manufactured. A complete life-cycle assessment would consider the life span of each type of shoe, in case one would stand up to regular wearing better than the other. Finally, then, the options for disposal of each type of shoe and the potential impact of each would be considered. Can the materials be reused or recycled? If not, what are the disposal options and the potential long-term effects of disposal? Conducting a comparative life-cycle assessment will not necessarily make such a purchase decision easy. Life-cycle assessment does not address things like animal welfare concerns among those who eschew leather products. Important trade-offs are implicit in such a purchase decision. On the one hand, cattle production as a source of leather (generally hides are a secondary product in the meat-production industry) is also a source of methane, a significant greenhouse gas. On the other hand, leather is a renewable resource, while crude oil is not. The host of chemicals used to process leather differ from those used to produce faux leather, and the environmental impacts of the various chemicals will be different, which means concerns about water pollution may have to be weighed against concerns about air pollution. In the final chapter of this book we visit some of the really challenging questions people find themselves asking about how our choices affect sustainability.

A relatively recent extension of life-cycle analysis is the idea of footprint assessments. Two common examples are carbon footprints and water footprints. While a life-cycle assessment should in principle consider all ways in which production, use, and disposal of products affect the environment, footprint analyses focus on dimensions of the environment, rather than ingredients or components of a product. They provide options for broader and more comparative assessments that might be made by policymakers. Calculating a carbon footprint for a product or an activity shows the amount of greenhouse gases generated. An individual can, for example, determine their personal annual carbon footprint by assessing things like the type and amount of energy used in their home; amount and modes of travel; types and amounts of food purchased, prepared, and consumed; purchase, use, and disposal of other consumer products; and use of or participation in many types of services or activities.[2] Companies are using carbon footprinting as a way to manage greenhouse gas emissions (so they are using carbon footprint as an indicator). Similarly, a product's water footprint describes the direct and indirect ways in which water resources are used in its production and consumption. Working with The Nature Conservancy, the Coca-Cola Company has been a leader among industries assessing how much water—and from where—their operations and products use worldwide.[3]

Are environmental indicators controversial?

Oh boy, are they! The stocks, flows, and feedbacks measured through indicators determine which aspects of environmental quality are monitored. People trained in environmental science agree on the relative strengths and weaknesses of different indicators for monitoring environmental quality, but these professionals are also aware that they work in a context where there is little agreement on the goals or rationale for environmental protection. Interest groups differ over which element of environmental quality should be prioritized. Early efforts to maintain a sustainable yield (discussed in chapter 3) for wildlife populations emphasized the stocks and flows of animals that were fished and hunted. Some fishers and hunters were in it for sport, while others were earning a livelihood. Other groups wanted to visit wilderness habitat for hiking or camping and still more wanted wilderness protected from all forms of human impact. The indicators for sustainable yield provided little insight into these preservationist goals. The wilderness habitat of game species was also a place where still other groups wanted to cut timber, raise cattle, or engage in mining. The choice of which indicators to monitor can favor some interest groups while pushing the things that matter to another group into the background.

Environmental quality professionals negotiate these interest group conflicts within a context buffeted by additional complications. Conflicting ideological views on the role of government vis-à-vis market forces and private control dictate different understandings of who (if anyone) should decide which features of environmental quality should be tracked by indicators. Even when there is consensus on the role of government, one can dispute whether authority should be vested in a local, regional, national, or even international agency. In addition, some aspects of environmental quality that a given group would like to protect are difficult to associate with reliable indicators. Some indicators are costly to monitor. Some forms of environmental protection may have existed since antiquity, but widespread appreciation of threats to environmental quality grew exponentially during the last half of the twentieth century. Whether operating as private landowners, corporate decision makers, or policymakers in government offices, environmental professionals struggle with the value conflicts that surround their selection and application of environmental indicators.

Does sustainability change how people understand and protect environmental quality?

Yes and no. We will start with no. If one takes the perspective of an environmental professional with job responsibilities that include monitoring or managing impact on environmental stocks and flows, there was no sudden moment when they started to think of their job in terms of sustainability. The tools they could work with changed gradually. Some of the tools came from ecology, discussed in chapter 3. Other tools came from environmental economics. For example, economists developed approaches for comparatively evaluating the value of different uses for a resource like wilderness. This allowed someone in charge of public lands to compare the value of an extractive use, like cutting timber, and a recreational or aesthetic use, like hiking, camping, or wilderness photography. Having a picture of multiple uses in hand, an environmental professional is in a position to start asking how changes in the stocks that matter for one use (say, a stand of virgin forest) affect flows to other stocks (say, tourists coming to camp). The ability to ask this kind of question is a major step toward thinking about environmental quality as a sustainability problem, but environmental professionals who took such steps did not necessarily think that they were doing something new.

However, when one steps back from the ground-level perspective of the individual professional, the accumulation of small changes in measuring, monitoring, and promoting environmental quality does add up to something bigger. As stocks, flows, and feedbacks are modeled, an environmental professional is able to look for points in the entire system of human practices and natural processes that makes up a forest, a watershed, a town, or even a planet where it is possible to initiate change. They can form hypotheses about which points in the system are most likely to move the indicators they are monitoring. For example, air pollution from factories could be changed by passing a law that limits what each factory emits. But putting factories out of business could have negative impacts on employment. It may be possible to have a larger systemic impact on environmental quality by creating an economic incentive to reduce pollution, while dampening the feedback to local employment. Policymakers who took steps toward this type of thinking may not have used the word sustainability to describe their approach, but the gradual growth of a systems orientation is what everyone needs to know about sustainability and environmental quality.

There were also some important conceptual innovations, new ways of thinking about protecting the environment. As just described, what matters about environmental quality differs for different people. There is no commonly agreed-upon definition of the problem. The science for developing indicators has progressed dramatically, but it is not perfect and there are many uncertainties. Even worse, decisions can have irreversible and costly consequences. As Horst Rittel and Melvin Webber put it, the decision maker has no right to be wrong. Environmental professionals have become more and more sensitive to these conflicted and poorly structured elements of the tasks they face. In chapter 8, we discuss how Rittel and Webber's idea of "wicked problems" helps people trained in sciences (where the answers are supposed to be cut and dried, objective, and free of biasing value judgments) cope with contingency and conflict. In the present case, the uncertainties and value conflicts concern which aspects of environmental quality they should monitor. The turn to sustainability requires more inclusive ways of addressing these conflicts and, most importantly, understanding people.

Like the evolution in systems thinking, recognition of the role that value conflicts play in protecting environmental quality has grown steadily. State or local officials responsible for environmental protection and corporate employees who monitor environmental impacts and work with both government and advocacy groups were among the first to understand. Though trained in the natural sciences, they now see that their job demands people skills. Again, they may not have conceived of this as a step toward sustainability thinking, and they would have been unlikely to use the word sustainability to describe it. Yet as the wicked nature of protecting the environment came to figure more prominently in the mindset of environmental managers, the profession edged steadily toward a sustainability framework. Indeed, the terminology of a "wicked" problem makes some scientists uncomfortable. Talking about environmental quality in terms of sustainability acknowledges the values-oriented aspects of environmental protection and avoids connotations of wickedness.

In short, the things many people wanted to protect and preserve in nature are now seen to be embedded within a system that can potentially connect to seemingly unrelated things. Tools that strengthen our understanding of those connections can help us identify the places where change in human practice will have the most leverage on the indicators chosen for monitoring. At the same time, the choice of which indicators to monitor is understood to be driven by values. Some of these values run deep: Is environmental quality limited to the values people associate with it (anthropocentrism), or should it include the value that other living creatures derive as well (biocentrism)? The turn toward sustainability has given people with narrow training in the sciences a way to cope with the conflicting values and points of view that impinge upon their responsibilities. It is less clear that a similar broadening of perspective has strongly influenced how nonspecialists approach environmental quality, but that is why we decided to write this book.

5

SUSTAINABLE DEVELOPMENT

What is sustainable development?

Whoa! Aren't we getting ahead of ourselves a bit? We should start with development, and once we're clear on that, we can go on to sustainable development. However, it can't hurt to make just a few observations before getting underway. First and foremost, many people seem to presume that sustainability and sustainable development are interchangeable. If you've been reading this book from cover to cover, you know that we do not agree, but if you've skipped directly to chapter 5, you may need a little coaching. We've answered the question "What is sustainability?" as follows: sustainability is a measure of whether (or to what extent) a practice or process can continue. Development is a process, so it is meaningful to ask whether and under what conditions it can continue. That is how we will (eventually) get around to answering this question.

Second, as we've hinted all along, much of the impetus for studying sustainability arose during the 1970s and 1980s when economists began to admit that economic expansion (or growth) was (or soon would be) constrained by industrial pollution and depletion of natural resources. The strong correlation between economic growth and the most common ways of measuring development meant these constraints were a challenge to the goal of economic development. Strategies for achieving development goals without violating these limits came to be characterized as sustainable development. This way of thinking was especially influential among policy-makers who were thinking about economic development on a global scale. To make sense of this influence and understand how it shapes thinking on sustainability, we must dive into the way policymakers think about development. We may dive more deeply than some readers want to follow, but getting a better handle on this domain of economics will be helpful for others.

This focus on economic development does not explain why sustainable development came to be identified with sustaina-bility itself. Maybe some people believe that promotion of ec-onomic development is so important that they just never get around to asking what sustainability could mean in connection with other activities (and practices, institutions, or systems of interaction). Alternately, some people may think that activities like running a business, producing food, governing a city or region, constructing and maintaining built infrastructure, or supplying people with energy, transportation, or health serv-ices can all be lumped together under some comprehensive idea. Maybe development *is* that idea. Although all this may sound abstract, these disparate activities cohere into a system of sorts. If you can make sense of that, then you can use the system concepts we discussed in previous chapters to judge whether and to what extent this conglomeration of practices and processes can continue to function and reproduce itself over time. With that thought, let's go back to the basic question.

What is development?

As a simple word or concept, development is defined in ways that vary according to context. Here are some common syn-onyms for development: evolution, growth, maturation, ex-pansion, enlargement, spread, progress. Many people think of development as something done by people in real estate—the business of buying, selling, and improving plots of land or houses or commercial buildings. Here, a developer purchases land or buildings and then invests more money in infrastructure (e.g., roads or a sewer system), in building structures (e.g., homes or shopping centers), or in repairing and renovating existing buildings. The developer hopes to sell or rent the property at a profit. Although this focus on real estate can be misleading, it is not a bad place to start. The broader notion of development that is relevant to sustainability is indeed a process where investments or expenditures are made with the goal of achieving an increase in the overall value of an asset, though it is usually understood to apply to society as a whole, rather than to the profit-seeking activity of an individual or a firm. We take some pains here to make this idea clear.

If we stick with the case of real estate for a moment, it is easy to get a handle on making investments that will increase the value of, say, a run-down house by more than the amount a developer puts into it. Television shows about "flipping" houses illustrate it quite well. The developer buys a house that nobody wants for $100,000, spends $30,000 to fix the plumbing, paint the walls, and replace the carpets, and then voila, she can sell the house for $200,000 (multiply these amounts by ten if you live in California). However, for getting to the idea of development that matters for sustainability, in addition to thinking about society as a whole, it is important to go beyond dollars and cents. The kinds of improvements that matter for development are improvements in the quality of life, rather than profit.

Here it is important to see the difference between an improvement that comes about through investment and one that comes about just by spending more of the resources you have on hand. Consider this scenario. You are troubled by the curb appeal of your home. Your landscaping leaves much to be desired. One option to improve things is to visit a garden center in the spring, buy lots of colorful annual flowers, and strategically plant them to make your front yard more attractive. Suppose this will cost you two hundred dollars. A second option also involves a visit to the garden center in the spring, but instead of buying annual flowers, you invest in shrubs and perennial flowers (the kind that will bloom every year—so long as you care for them). This requires an investment of five hundred dollars. In either case, you make an expenditure, and you are happy with your landscaping either way. The first option is just consumption, though; you've spent money to make yourself happier. The second option increases your capacity to experience happiness on an ongoing basis. You will enjoy your landscaping the next year, and the year after that.

Two points are important in this discussion of development. One is that the growth, expansion, or enlargement we are associating with development is not an increase in happiness, though some economists might say that it is. The economic theory of development would characterize this as an increase in welfare. Of course, the person who flips a house for a $70,000 profit is probably happier for doing that too. In both cases, an economist would say that there has been an increase in welfare. However, it *is* important that the improvement or enlargement concerns quality of life—that's what an economist means by associating welfare and happiness. The second point to notice is the difference between consumption and investment. Consumption is any expenditure where the benefit—the happiness or welfare—that you achieve is roughly correlated with what you give up to get it. Investments are expenditures that increase the capacity or capability for production or continuous generation of welfare. Development occurs when the capacity for continuous generation of welfare is enhanced. Investment is closely associated with the economic concept of capital, so there are two concepts that are crucial to understanding development: welfare and capital. Development is an increase in welfare and capital, but these two notions must be properly understood.

What is welfare?

To begin with, we are not talking about a payment or check that someone receives from the government because they are unemployed, indigent, or injured. Welfare is a general concept that has long been used by social theorists to indicate how well a person or group is faring in life. It is a composite of other familiar ideas. One is health: disease, disability, and physical distress are detriments to one's welfare, while vitality, vigor, and physical ability are enhancements. Welfare is also associated with contentment, happiness, satisfaction, and other components of one's mental outlook. More broadly, a person's welfare is greater if they have meaningful choices and some degree of control over how their lives (and the lives of people they care about) go. A sense of belonging to a community or affinity group is a component of welfare, while alienation, loneliness, or isolation are generally detrimental to a person's welfare. There is, in fact, room for an entire book called "Welfare: What Everyone Needs to Know," but you are reading a different book, so we will cut this short. When we define development as growth in general welfare, it means an increase in the total amount of happiness, satisfaction, or wellbeing of all persons in a given group. Who gets included in said group? In theory, everyone. There are serious arguments that "the general welfare" even includes animals. For practical purposes, estimates of welfare are usually made for populations of people who are residents in a given political entity such as a country, a region, a state, or a city. As an idea, development means improvement or expansion of welfare among a group. However, there are important details. Improvement or expansion of *group welfare* means that total welfare can improve even when some people are made worse off. If more people are being made happy or healthy or have expanding opportunities even while a smaller number of people are seeing a decline in their welfare, you still have development. What is more, approaches to measuring an increase in welfare

are also open to dispute (as our readers will soon see). Wealth and poverty refer to the general state of an individual's welfare or to the aggregate welfare of some population or group. Quantifying all this is a complicated business, but increasing total welfare through increasing stocks of capital is what motivates the theory of sustainable development. Some inequality in the distribution of wealth may be inevitable or even desirable, but extreme inequalities are associated with numerous social and moral issues. You can find answers to more of your questions about the distribution of wealth and poverty in the book *Inequality: What Everyone Needs to Know®*, by James Galbraith.

What is capital?

We introduced the idea of capital in connection with an investment that increases the capacity for producing welfare. More generally, any resource, material good, input, or factor that contributes to a production process while not being consumed or used up in the course of that process is classified as capital. This is the core idea of capital in classical and neoclassical economic theory. Capital has been redefined and re-conceptualized in other contexts to suit a wide variety of explanatory and persuasive purposes, but we are sticking to this definition of capital because it is used in explaining sustainable development.

You can understand the concept of capital by thinking about the production of some ordinary physical goods: a bicycle, a carton of orange juice, or a cell phone, for example. Physical resources (metal, oranges, oil for making plastic) go into these things, and these physical resources are going to be consumed or transformed in some way as you make the bicycle, juice, or phone. The labor time of the people who make them is consumed in the course of production. But the factory where these are made or the knowledge and skill of the people who make them are not used up. They are still around to make more. These elements that are not consumed or transformed in the process of production are called capital.

Although capital endures in the process of manufacturing goods like bicycles, orange juice, or cell phones, it can nonetheless be understood as a stock that increases or decreases. (Note: if you skipped ahead, we covered stocks and flows in chapter 3.) How can anyone make sense of capital increasing or decreasing? This is simple in concept. If you build a more efficient factory or improve the skills of those making the bicycles, you increase capital. Obviously, the amount of consumable resources used is going to increase as you produce more goods (each bicycle you make is going to require a certain amount of metal, plastic, rubber, and whatnot). But investing in better machines or increasing the skills of workers could mean that you are able to increase the number of bicycles produced without a proportional increase in the materials used.

How does capital decrease? Economists use the word depreciation. Sticking with bicycle factories, you might notice that they deteriorate over time. The factories and skills for making something can also become obsolete. Consider a traditional blacksmith or wheelwright shop, equipped with capital (e.g., tools and skills) for making bicycles that someone might have seen at the shop of Wilbur and Orville Wright in Dayton, Ohio, around 1900. The tools and skills endure, but with changes in technology like assembly lines and interchangeable parts, they become obsolete. In fact, the whole bicycle shop becomes obsolete when people stop buying bicycles because they can buy inexpensive automobiles. This bicycle versus car example illustrates that when economists talk about a growth or decline in capital, they are imagining a quantity that is measured in reference to an economy-wide system of exchange. This is also a point that will become important below.

However, let's not forget that when we talk about development, we're not talking about making bicycles, cartons of orange juice, or cell phones. We are talking about the welfare (i.e., the happiness, satisfaction, health, or well-being) that is produced when people *have* bicycles, orange juice, or cell phones. As long as people are out there spending hard-earned money to buy bicycles, orange juice, and cell phones, it is reasonable to think they are doing so because it generates some kind of satisfaction for them or makes some improvement in the quality of their lives. So as we move from production economics to welfare economics, we tweak the notion of capital a bit: now we are talking about the economy-wide capacity of society to produce welfare. But many points carry over.

Notice that our examples of capital for producing these goods include both physical infrastructure (factories and technology) and intangibles like skill. This will be an important point to pick up on later. Generalizing from the bicycle production example, where we lumped together all the various ways one might invest in capital to produce more of them, we can lump together all the ways in which capital investments can increase all productive activities in society to make more of the things that people want. The take-home point so far is that if you want to encourage development (an increase in general welfare), a key means of doing this is to encourage the growth of capital.

If we string these ideas together, one criterion for the sustainability of development would be whether (or to what extent) capital can continue to grow. If we think of capital and welfare as stocks, there will be feedback from the capital stock to the inflow of welfare. That is, as capital grows, it stimulates the flow that causes an increase in welfare. Stock and flow systems thinking allows us to describe development as a system that is sustainable as long as the capital stock is increasing but is threatened or unsustainable if inflows to and outflows from the capital stock are vulnerable to shocks, oscillation, or collapse. This way of thinking allows us to understand the sustainability of an economic system much like ecologists understand the sustainability of ecosystems. Of course, we haven't said what drives growth in the capital stock or what could threaten its flows. That is what the rest of this chapter is about, and to get there we need a few more economic concepts.

What limits the sustainability of development?

There is a naïve answer and a more sophisticated answer. The naïve answer is not wrong, and we will give it now, saving the more sophisticated answer for when we have considered a few more questions about the theory of economic development applied on the global scale. In short, making bicycles, orange juice, or cell phones requires (in addition to labor and capital) all the stuff, the metals and other materials, that goes into them, along with the energy needed to move the stuff around and power any equipment you are using. Manufacturers make more and more of these things as the population grows and consumption increases. But the earth is finite. People can't go on making bicycles, orange juice, and cell phones forever, because sooner or later human beings will run out of the resources they need to make them—metals, fertilizers, and oil, for example.

On top of that, we've learned that making bicycles, orange juice, and cell phones produces some stuff nobody wants: various forms of pollution that damage health and degrade the planet. If people keep on producing that pollution, they won't have breathable air or drinkable water. These harmful outputs from production processes can at least theoretically accumulate to the point that our planet is no longer habitable. Running out of materials and the accumulation of pollution together place a limit on the number of things people can make. If having those things is important for welfare, one can infer that limits on our productive capacity limit the potential for increasing welfare. Since increasing welfare is what development is all about, it stands to reason that there are limits on how long this process of development can continue.

This is all correct, and this idea of "running out of stuff" may in fact be precisely the way that many people actually do understand sustainability. But economists have argued that as people do start to run out of the materials that go into bicycles, orange juice, and cell phones, the prices of those products will go up. And as the prices go up, people find different ways to spend their money. Are they just as happy or satisfied as if they had acquired bicycles, orange juice, or cell phones? That's a question we're not prepared to answer, but so long as they are spending their money on something, the economy will continue to hum along. People will have jobs, and the machinery and skills needed to make those other things can continue to accumulate. So there is a sense in which running out of the stuff manufacturers need to make bicycles, orange juice, or cell phones doesn't necessarily threaten development so long as society (that is, producers and consumers alike) isn't depleting capital; it just drives us toward a different pattern of consumption. Are the economists who make this argument right? Or are they just blowing smoke? Thinking back to chapter 4, how do pollution and resource depletion affect capital stocks? In order to answer these questions, you need to understand how economists measure economic growth. You also need to recognize that concerns about natural capital stocks, which we will discuss later, add additional layers of complexity to the problem. These complexities get us to a more sophisticated answer.

What is economic growth?

Economic growth is the increase in an economy's capacity to produce goods and services. The assumption that access to more goods and services improves welfare is fundamental in economics, and since economists cannot measure welfare directly, they measure the production of goods and services. Welfare can go up or down for any given individual as their health, happiness, contentment, or capability increases and decreases, but most of us would have a really hard time making a numerical estimate of our own welfare at any given moment. Making an estimate of the total welfare for everyone would be even harder. (It is, in fact, mathematically impossible to derive a social welfare function from measures of individual welfare.) Although theorists continue to debate the concept and measures of welfare, focusing on an observable quantity is more practical. Recall the argument that someone spends money to buy a bicycle, carton of orange juice, or cell phone because it creates satisfaction or improves their quality of life (i.e., welfare). Thus, purchases of goods and services can be used as an indicator of welfare. At least, that's how one very influential view goes. There are rather serious flaws with this argument, as we will show. (Note, we discussed types of indicators in chapter 4.)

One can think of a region's or nation's economy as a massive system of stocks and flows. This system can be analyzed by examining sectors: agriculture, mining, manufacturing, transportation, healthcare, energy production, and so on. Firms in each sector are producing goods and/or services and selling them either to consumers or to other firms (often in other sectors). Our bicycle maker will sell bicycles, but to make them she will be buying metal goods and energy from other firms. You get a sense of the bicycle manufacturer's economic well-being (their sustainability, if we go back to chapter 2) by looking at their balance sheet. But to get a sense of the entire economy's "bottom line," you would have to total up exchanges of all firms. That's daunting, but getting a handle on each sector's buying and selling is doable. In fact, economists have been doing it for the US economy since the 1930s. The economists who estimate Gross Domestic Product (GDP) on a quarterly basis are adding up the buying and selling across all sectors in the economy. Although economic growth refers conceptually to any expansion of economic activity, it is typically measured simply in terms of an increase in GDP from one quarter to the next. If GDP goes down, the economy is shrinking. If an economy shrinks continually for longer than six months, that is described as a recession. Economists also look at the status of several other indicators like unemployment and real income to diagnose a recession, but we don't think everyone needs to know about these details to understand sustainable development.

GDP and the concepts of economic expansion and contraction were not originally created to explain or measure capital stocks or the capacity to produce welfare. They nonetheless serve as a reasonable proxy. As a technical economic notion, then, economic development is understood to track closely with GDP. The primary caveat is that while GDP can fluctuate periodically in conjunction with cycles in the activity of individual firms (i.e., the business cycle), economic development refers to longer-term expansion and accumulation of a society's overall capacity to produce a high quality of life—a high standard of welfare—for its citizens. We'll come back to the role of GDP in measuring development later.

Historically, economists and other social theorists presumed that although a nation's economy (and occasionally the global economy) would go through cycles of expansion and contraction, increases in aggregate social welfare are the overall pattern. This is to say that development reflects a steady upward trend in economic growth, even as there may be periods of contraction. The long-term upward trend of development was observed to eventually overcome even extended periods of economic contractions in economic activity, or depressions. By the 1950s, economists observed growth in GDP over a long period for industrialized countries in Europe and North America, as well as Japan, Australia, and a few others. But this growth did not occur everywhere, and the uneven pattern of economic expansion and associated disparities in wealth and quality of life were exactly what led to both theoretical and practical efforts to stimulate development in less industrialized parts of the world. These included former European colonies in Africa and Asia that attained independence following World War II as well as countries that relied heavily on resource extraction and export. This disparity was central to the work of the Brundtland Commission (discussed in chapter 1) and the formulation of their famous definition of sustainable development.

How are wealth and poverty related to welfare and economic growth?

Defining and measuring wealth and poverty are topics within development economics, which is itself a field within welfare economics, where the characterization and measurement of changes in welfare have seen a long and complex history. This body of theory lies behind some of the debates over sustainable development discussed later in the chapter. We can simplify the various approaches to defining wealth and poverty for present purposes by holding fast to the description of welfare that we have already given. Wealth and poverty refer to the general state of an individual's welfare or to the aggregate welfare of some population or group. Poverty is a concept for recognizing deficits in wealth such that one's welfare falls below some level that is determined by ethics and social norms. One way of indicating poverty is a lack of basic necessities such as food, water, shelter, and clothing. Extreme or absolute poverty occurs when people are deprived of these basic needs. The current World Bank standard for extreme poverty is an income of less than $1.90 (US dollars) per day.

Previously, we noted that the welfare of individuals increases when their ability to purchase goods and services increases. We also discussed how growth in GDP (which occurs when purchases of goods and services increase) is used as a proxy for improvements in the aggregate welfare of individuals living in a given region or country. But this measure is not sensitive to the fact that wealth is not evenly distributed among those in the region, and so welfare is not evenly distributed either. The Brundtland Commission was charged with evaluating global development. The per capita wealth in industrialized regions (the United States, Canada, Japan, Australia, and Europe) is much greater than in Africa, South

America, and much of Asia (though China has grown rapidly since 1987). These inequalities are associated with social and moral issues because they result from exploitive histories of slavery, racism, and colonialism. As readers will see in the next chapter on social justice, inequality and exploitation also create direct problems for sustainability. In the rest of this chapter we emphasize how a systems approach to understanding links between pollution, resource depletion, and global development challenges some of the fundamental assumptions in development theory.

What is global development?

Not surprisingly, global development is simply an application of the concepts discussed throughout this chapter on a global scale. But this is not as straightforward as it might initially seem. Grasping issues at a global scale is key for moving from a naïve understanding of sustainable development to a more sophisticated understanding. The historical starting point is the creation of international organizations such as the United Nations (UN), the World Bank, and the International Monetary Fund (IMF) at the end of World War II. At that juncture, world leaders began to see the vast inequalities between industrial countries and the less-industrialized nations of the global South as a problem.

Debates over the way in which economic growth in regions with high poverty rates could be stimulated and how the welfare of their people could be improved shaped development theory significantly. The tensions between socialist economies (such as the Soviet Union) and Western states was an important part of this story. This is because conflict between the United States and the Soviet Union during the Cold War era (1947–1991) often took the form of skirmishes within and among states and resistance groups in these less industrialized nations. Would capital growth and improvements in welfare be achieved through government intervention or through private investments? Although the details of this debate lie beyond the scope of this book, it is important to bear in mind how alternative approaches to development have been influenced by larger sociopolitical debates over socialism, capitalism, and the contours of geopolitical relations.

What is most significant for understanding sustainability is that all conceptions of development (socialist or capitalist) initially envisioned that the wealth gap between the industrialized world and the global South would be closed as economies in less industrialized states caught up. This was both a political and a moral problem, since the low levels of capital in less industrialized states were the legacy of colonialism. The wealthy countries were seen as morally responsible for the poverty of former colonies because they became wealthy through exploiting the labor and resources of colonized places. Closing the wealth gap was a political problem because trade relations and foreign investment (possibly in the form of development assistance) had to be negotiated through global institutions like the World Bank, the IMF, and, eventually, the World Trade Organization (WTO).

Development and economic expansion in these less industrialized former colonies—whether achieved through government programs or private initiatives—were viewed as solutions to these moral and political problems. However, this vision was on a collision course with the limits to development reflected in the exhaustion of natural resources such as oil and other minerals, ocean fisheries, or fertile croplands on the one hand and the effects of industrial pollution on air and water on the other. Discussion of what we've called the naïve appreciation of limits to development began among global governance institutions associated with the UN at a 1972 conference in Stockholm.

Within the context of sharp inequalities in both economic activity (as measured by GDP) and quality of life, this problem looked even more complex at a global scale. Highly developed economies such as the United States were consuming a disproportionate share of the world's resources and emitting a disproportionate amount of the world's pollution. Putting these inequalities in the context of environmental impacts made the problem more complex because countries with high rates of poverty could not catch up just by consuming more resources and emitting more pollution. The calculations of global capacity indicated that the planetary ecosystem would simply collapse if that were the case. (In fact, this bleak picture has only grown darker as the impacts of climate-forcing pollution have been added to the equation.) What was to be done?

The Brundtland Commission (officially, the World Commission on Environment and Development, or WCED) was convened in the 1980s to address the moral, economic, and political dimensions of this situation. The options were stark. Advocates for impoverished nations were adamantly opposed to the suggestion that they should forgo expansion of their economies and a corresponding improvement in their quality of life. Although some saw the moral case for retrenchment of economic activity in the industrialized world in compelling terms, it has always been viewed as politically unacceptable. Given this impasse, the Brundtland Commission came up with its definition of sustainable development as "development that meets the needs of the present without compromising the ability of future generations to meet their own needs." The high profile of the WCED pushed the language of sustainability to the forefront, and the publication of their report *Our Common Future* in 1987 launched a new era of theorizing and debate over whether this vision is practically achievable.

How can sustainable global development be achieved?

The answer to this question hides in the details of everything we have written so far. Some of the details reside in the general theory of development or in the way that GDP has been used to measure it. If GDP is not a reliable indicator (see our discussion of indicators in chapter 4), it may not be steering us toward improved welfare. The world won't achieve sustainable global development if development specialists are tracking an unreliable indicator. The deeper significance of using GDP is embedded in debates over the strategy for continuing economic growth through expansion of capital. Answering whether sustainable global development is achievable requires us to return to a more sophisticated understanding of what limits development.

Our naïve account of limits is not wrong in noting that stocks of many natural resources are finite, while the pollution associated with many industrial practices (including climate-forcing emissions) is adversely affecting human welfare. But development involves an increase in capital, and this is not the same thing as an increase in the production and consumption of goods that use up resources or emit pollution. Even if the consumption of finite resources and the depletion of environmental quality have to stop, it is at least theoretically possible for development (meaning growth in welfare, which economists equate with growth in total economic activity) to continue. Perhaps it can even continue indefinitely. That was the thought behind an influential lecture entitled "An Almost Practical Step toward Sustainability," given by Robert Solow in the early 1990s.

One devil that resides in these details takes us back to the definition of capital. Any factor or input in production that endures throughout the process of production can be categorized as capital, and our examples included both built or constructed facilities (factories and machinery) and the skills and knowledge that it takes to make something like a bicycle. It should be obvious that since factories and machinery are themselves made from finite natural resources, there cannot literally be continuous expansion of capital in that particular form. This points us toward other, nonmaterial forms of capital, like knowledge and skill. Development might be sustainable if people are able to shift more and more of the growth in economic activity to reliance on forms of capital stocks that do not themselves involve the depletion of finite resources or the degradation of environmental quality. And that leads to our next question.

What are the forms of capital?

The factories, tools, and machinery that are used in a production process constitute *built* or *manufactured capital*. We have already mentioned the contribution of skill, knowledge, or ability to production; that is *human capital*. The idea here is that as people become smarter, they can economize on other factors that are used in a production process. One can envision human capital as consisting of a given individual's experience, wisdom, or knowledge, or more generally as the accumulated knowledge possessed by society at large. In either view, education plays a key role in building human capital.

In chapter 2, we discussed social permission to operate and linked it to a firm's reputation within a community. A firm with a good reputation is said to have social capital. A bit more nebulous than skill or knowledge, the idea of social capital encompasses productive capabilities that emerge out of a social group's capacity for cooperation and collective action. This may consist of trust or goodwill that is tied to reputation, but it can also take the form of norms and traditions that allow people to engage in productive activity at a lower cost than they might otherwise incur. In some cases, social capital enables productive activities that would be entirely impossible without cooperation. The tradition of barn raising, where neighbors in rural areas pitch in to help with complex construction projects that could not be achieved by a farm family working on its own, provides an illustration of social capital.

Business schools will emphasize *financial capital*, which is simply money that is treated as a base stock for supporting business activity. Money flows in and out of a business as it buys supplies, pays wages, and collects revenues. The financial capital stock is the base amount that is not used up in the sense that it remains constant (and hopefully grows some) after all of these transactions clear. *Venture capital* is a form of financial capital that represents the base money stock that an investor spreads around to various (often high-risk) firms. The expectation is that after all the profits and losses have been tabulated, this basic stock is reproduced (it is not "used up"), and the investor takes some profit as income. But these are all examples that are somewhat peripheral to sustainability. They simply illustrate how someone might think of growing capital even in the absence of doing something that draws down finite resources or contributes to environmental degradation. The form of capital that has been central to debates about sustainable development is *natural capital.*

What is natural capital?

There is no firm consensus on the answer to this question. The general idea is that nature itself produces a lot of things that contribute to human welfare. This is not the same thing as supplying natural resources or consumable goods, like oil and gas or fish from the sea. When people run short of resources or consumable goods, prices of those things increase, and they either economize on their use of them or shift to using something else. Natural capital is more like nature's capacity to produce goods and services that are important to us as opposed to the goods themselves that human beings take from nature. Beyond the ocean's capacity to produce fish, natural ecosystems do a lot of work for us in the form of making it rain, filtering the air and water, regulating our exposure to solar radiation, and many other examples. The Millennium Ecosystem Assessment described four ways in which natural capital provides important goods and services: direct provision of goods and services, regulation of natural processes, support of cultural values, and the ways in which different natural systems support one another. (We answered the question "What are ecosystem services?" back in chapter 4.)

The general idea of capital came from nineteenth-century economists. They may have taken the idea that natural processes also contribute to productivity for granted, but in any event, this idea did not formally enter into thinking on economic development until recently. Again, it's important to understand that the idea of natural capital relates more to natural capacities for producing welfare than it does to the stocks of natural resources or to the stocks of renewable resources that are harvested on an annual basis. The theory of *ecological economics* arose in the 1980s in response to questions about whether capacities that produce ecosystem services can be degraded or damaged in ways that reduce that stock of natural capital. Although most theorists who have studied these questions agree that they can, whether this is a problem is a debate that rages on. Part of that debate is about whether world leaders and economic development specialists pursue weak or strong sustainability.

What are weak and strong sustainability?

Weak sustainability is the view that declines in natural capital can be offset by gains in other forms of capital. In this view, you can still have sustainable development even when nature's contribution to welfare is declining. For proponents of weak sustainability, all the various types of capital are fungible, or interchangeable, with respect to one another. In particular, if nature's capacity to produce goods that contribute to welfare degrades or is damaged, the increase in technical capacities, human skills, and financial wealth that comes about through economic development can more than make up for it. Total capital (and total welfare) can therefore continue to increase (or grow) indefinitely. As a very simple example, coastal wetlands in some regions have been replaced by levees and sea walls. The levees and walls help prevent coastal flooding, which is what the wetlands did before they were eliminated.

Strong sustainability is the view that substitutability among the various forms of capital is limited; that is, some forms of natural capital cannot be replaced by growth in technical abilities or the accumulation of other forms of wealth. Strong sustainability also argues that some things economists have traditionally thought of as contributing to development (like manufacturing that releases air and water pollutants) are not actually helping. The loss of natural capacities, according to this view, is leading to an irreversible decrease in welfare, which implies that modern industrialized societies are actually losing capital, rather than increasing it. Proponents of strong sustainability argue that loss of coastal wetlands and their ability to buffer storm surges also comes with a concomitant loss of species habitat and recreational space—both of which are types of natural capital. Increasing hurricane risks due to a changing climate also represent a loss of natural capital. And when the levee breaks, the ecosystem services provided by coastal wetlands are sorely missed.

It is worth noting that one's views can align with weak or strong sustainability to varying degrees. One might think that some substitution is possible but that eventually some substitutions will have to stop. This is the position taken by Herman Daly, one of the economists who helped the field of ecological economics get a start. One might also think that certain forms of natural capital really are essential, while others are not. One example of an important but (perhaps) inessential type of natural capital might be natural areas that contribute to welfare by providing for recreational or spiritual experiences. These contributions are important, but perhaps people in the future can learn to do without them. (Of course, not everyone will agree.)

Even among people who do not really think deeply about the various forms of capital and their role in economic development, the divide between weak sustainability and strong sustainability still plays an important role. People who think that humanity can innovate its way out of the global predicament are said to embrace weak sustainability, while those who think that people have to constrain their impact on natural systems and shift to a very different way of life line up with the advocates of strong sustainability. We are not going to settle all the debates around weak and strong sustainability here. Regardless of your position in this debate, you need to consider what indicators like GDP tell us about progress toward sustainable development. And the topic of sustainability indicators opens a whole new debate: Is GDP a good indicator? Your response to that question is going to be determined in part by what you think about the substitutability of different forms of capital—especially whether increases in technical capacities, human skills, and financial wealth substitute for natural capital.

What is wrong with GDP as an indicator of sustainability?

Technically, GDP is an indicator of economic activity: that is what it was designed for. It is computed by adding up the dollar value of all goods and services produced in an economy in a given time period.[1] Calculating GDP allows us to estimate what each sector of the economy is contributing to overall productivity. As we noted earlier, so long as the total productivity (that is, GDP) of the economy is growing, policymakers have presumed that welfare is increasing. But many things that contribute to economic productivity have a questionable contribution to welfare.

Recall our earlier point that there are serious flaws in the assumption that increases in the buying and selling of goods and services represent increases in welfare. Consider expenditures on healthcare, one of the fastest growing sectors of modern industrial economies. On the one hand, spending on healthcare may indeed reflect an increase in wellness, and that would be a contribution to welfare. On the other hand, people may be spending more on healthcare because more and more of them are sick, or they are sick with more debilitating (and expensive) conditions. Arguably, reducing the need to spend money on healthcare in the first place could increase welfare.

There are other examples. If a society spends more money on policing or military defense because hunger or poverty (whether at home or elsewhere) are leading to social unrest, one may see an increase in GDP from those sectors of the economy. But again, whether these increases are really making the members of that society better off is questionable, particularly if one compares those expenditures with what might have been spent to prevent unrest in the first place. Similarly, if a coastal area is ravaged by a hurricane, the purchases of everything needed to repair and rebuild homes, businesses, and infrastructure boost GDP. But does that increase in GDP indicate that the residents of that coastal area are better off?

In short, a healthy debate continues about whether GDP is a very good indicator of welfare in the first place. If it's not, then measuring sustainable development in terms of continuous growth in GDP is probably not a very good approach. But it may also be a poor choice because of how it accounts or fails to account for capital. Economists include investment in some types of capital when calculating GDP. Purchases of new factories or equipment by a business are examples of investments in manufactured capital that get counted in GDP. One might argue that expenditures for education could, if studied carefully, indicate investments in human capital. GDP measures savings, and that is financial capital. But two important things don't get accounted for: investments in social capital and investments in natural capital. And if we are right and environmental quality (chapter 4) and social justice (chapter 6) matter for sustainable development, then GDP is leaving out something important. For those comfortable with weak sustainability, this may not matter—capital stocks are fungible. If people are investing in some type of capital, we're good.

Of course, we've already noted that the devil is in the details. Proponents of strong sustainability worry specifically about investments in natural capital. This may or may not show up in GDP. More precisely, they worry about the loss of natural capital—depreciation or disinvestment, if you will. The loss of coastal wetlands, and the ecosystem services they provide, doesn't show up in GDP calculations (although the expenditures to recover from hurricanes do). To be fair, GDP doesn't account for any depreciation of capital. Economists also calculate Net Domestic Product (NDP), which subtracts depreciation of manufactured capital like buildings and machinery from GDP. But how do you account for depreciation of human capital? And if social capital and natural capital aren't accounted for in the first place, NDP doesn't help. Proponents of strong sustainability are understandably critical of GDP as an indicator of sustainable development.

Finally, let's return to the inequality issue briefly. Comparisons of GDP across countries have to consider things like population size, since a very small country is not likely to achieve (nor does it need) as much economic activity as a much larger one. This is generally accounted for by calculating GDP per capita. Even this effort to refine the use of GDP is challenged, though, because neither GDP nor GDP per capita tells us anything about the distribution of wealth (welfare) within a given population. Following the Great Recession of 2008–2009 in the United States, critics of income inequality created the label "one-percenters" to refer to the fact that so much of the wealth in the United States is concentrated in the top 1 percent of the population. The Brookings Institution reported that in 2016 (data used for their 2019 report), the top 1 percent of the income distribution in the United States held 29 percent of the country's wealth, while the top 20 percent held 77 percent of the nation's wealth. In many countries around the world, wealth inequality is even more extreme. If growth in GDP, or even GDP per capita, masks such distributional imbalances, can anyone be sure it gives us an accurate picture of whether global societies are achieving any progress on the sustainable development front?

This is, in truth, a debate within the theory of development, rather than something that is crucial to sustainability as such.

Simon Kuznets, the economist who came up with the method for calculating GDP, cautioned from the start that GDP should not be considered a measure of standard of living, but almost immediately that was how it was used. Debates about sustainable development have reinvigorated criticisms of GDP that date back to the 1930s, when Kuznets did his work. In response, some theorists of sustainable development have come up with alternative ways to measure increases or decreases in social welfare. One of these is the *Genuine Progress Indicator*.

What is the Genuine Progress Indicator?

The Genuine Progress Indicator (GPI) is similar to the GDP, in that it measures economic activity. But the GPI is intended to assess many other variables that are presumed to contribute to well-being or happiness in addition to just production of goods and services. For example, it modifies the standard calculations of economic activity by adjusting downward as income inequality increases, reflecting the assumption that greater income inequality results in lower social welfare. It accounts for contributions to welfare that aren't reflected in GDP, such as the value of volunteer work. It adjusts downward for things like pollution, loss of natural capital, and crime—all of which are said to reduce well-being. Remember, cleaning up pollution, depleting natural capital, and fighting or recovering from crime all involve expenditures that increase the GDP.

The GPI (or a variation of the GPI) has been calculated for several US states and a number of countries around the world. Maryland and Vermont, Canada, and some European Union countries consider these calculations in budgeting and other legislative decisions. In all areas where GPI has been calculated, a singular message rings through. Comparing GDP changes and GPI changes over specific periods of time shows that during periods of robust GDP growth, GPI was stagnant or declining. This, it is argued, is evidence that GDP is overestimating the extent to which economic growth means real improvements in quality of life.

GPI and similar indicators are not without their critics. One principal argument goes like this: Adding up the value of goods and services produced is straightforward and consistent. For example, the dollar value of bicycles produced adds to GDP in a straightforward way because, product differentiation notwithstanding, bicycles and money spent for bicycles are a known entity. The same cannot be said for some of the elements that are included in GPI. For example, the value for every hour of volunteer work may not be the same. People do all kinds of volunteer work, and they may not contribute equally to social welfare. Similarly, is there some commensurable way to capture the costs of lost wetlands and air pollution so that such measures can be combined into a single indicator? A final argument questions whether any indicator can truly measure the well-being of a diverse society whose members have very different views of what is important and what adds to or diminishes social welfare. Clearly, measuring some of the items included in the GPI is challenging, but proponents of its use believe society as a whole benefits from making the effort to obtain a clearer accounting of general welfare.

Is sustainable development just equivalent to sustainability?

We think not. Our discussion of sustainable development has emphasized the way that the most quoted definition (meeting the needs of the present generation without compromising future generations' ability to meet their needs) emerged out of a very focused understanding of what development is and why it matters in a global context. We also think that sloppy thinking on this point (usually combined with skimpy understanding of what development is and the history that led to the Brundtland Commission) is one of the reasons why sustainability seems so mysterious and hard to pin down for many people. It is meaningful to ask whether a given development

trajectory is sustainable, and it is at least theoretically interesting to ponder the question of whether capital itself can
grow indefinitely.

It's also true that the "meeting needs" definition can be
meaningfully applied to many other human activities. Someone
could say that sustainable agriculture meets the needs of the
present generation without compromising future generations'
ability to meet their needs, and one might say similar things
about architecture, urban infrastructure, or transportation systems. It is less clear that this aphorism clarifies what it means
with respect to ecosystems or with respect to many of the other
environmental goods that were discussed in chapter 4. It's not
even clear that a sustainable business is one that meets the
needs of present generations without compromising future
generations' ability to meet their own needs. Business is often
much more about satisfying wants than meeting needs.

When one adds the concerns about GDP into the mix, the debate becomes even more complex. But perhaps the root issue is
whether one thinks that increasing the total stock of social welfare is itself what people should be trying to achieve through
sustainability. As we've seen, total welfare can increase even as
some people's welfare goes down. What if some groups in society are constantly made worse off even as total welfare goes
up? Is that really something people want to sustain? Recall
our discussion in chapter 1 that not everything that seems to
continue indefinitely aligns with the social goals embedded in
broader discussions of sustainability.

Then we can add in the possibility that humanity has moral
responsibilities to take the interests of animals into consideration, or to think of themselves as caring for the earth itself.
Some religious and cultural traditions have placed much more
emphasis on preserving and continuing natural and cultural
systems than they have on increasing the health, wealth, and
happiness of current generations. An understanding of sustainability that focuses on the continuous reproduction of social
practices or traditions or on the preservation of biodiversity

and the integrity of regional or planetary ecosystems might not presume that human welfare is the be all and end all of sustainability.

Again, we don't presume to settle these debates, but we do think that understanding what is at issue is part of what everyone needs to know about sustainability. Not everyone thinks that progressive and continuous development is what sustainability means. Simply presuming that sustainability means sustainable development is thus a barrier to the conversations that would contribute to a more sustainable society.

6

SUSTAINABILITY AND
SOCIAL JUSTICE

What is justice?

Justice is best defined by pointing out the distinct but
overlapping functions the idea of justice performs. Law enforcement, the courts, and prisons administer criminal justice: the system for arrest, trial, and punishment of those who
break the law. The word *justice* both names this system and
stands for the norms determining whether the criminal justice
system is functioning as it should. Someone accused of a crime
must be treated justly. When the accused is fairly tried, is con-
victed, and serves the sentence handed down by the courts,
people say that justice has been done. Here, the idea of justice
is being applied to the conduct of an offender (who has committed an injustice), to the police and the courts (who must
observe canons of justice in their treatment of suspects), and
to the result (justice is served when this system operates well).
We already see that the concept of justice is playing multiple
roles, and readers should not be surprised that debates and
philosophical discussions of justice date back to the earliest
periods in recorded history.

Even when the police, judges, and the penal system are
faithful to law, one can ask whether the law itself is just. In his
"Letter from a Birmingham Jail" Martin Luther King Jr. quoted
Saint Thomas Aquinas: an unjust law is no law at all. King was

incarcerated because he had violated laws mandating racial segregation. His appeal to justice called readers to evaluate the system of Jim Crow laws in the American South according to a moral standard beyond the law. In other words, any system of criminal justice is itself subject to review according to an ethical standard. King was calling upon Americans' sense of justice. As he later put it, "I have a dream that one day this nation will rise up and live out the true meaning of its creed: We hold these truths to be self-evident; that all men are created equal." Equality before the law was explicitly articulated as a principle of justice in the United States' founding documents, yet King was pointing out the inconsistency between principle and practice.

The idea of justice is also applied to more mundane matters. Conduct that falls below the threshold of civil and criminal law is also evaluated through the lens of justice. Norms of civility in daily interaction include allowing everyone an opportunity to speak in a discussion or standing in line to obtain service. A single violation of such norms might not be considered an injustice, but when social practices systematically discriminate in favor of one group or against another group, these patterns of interaction are judged similarly to the way that judgments are made in the criminal justice system. We're not saying that connecting justice and sustainability is a mundane matter. Our point is that we will be relying upon ways that the idea of justice figures in ordinary and familiar contexts as we discuss social justice and sustainability. In short, this chapter addresses the tension between what is and what should never be.

What is social justice?

Social justice is an ideal or goal that specifies what people living together in civil society owe to one another. Ideals of social justice highlight deficiencies in the configuration or performance of particular social institutions. The standards for social justice have varied over the course of human history. The literature of law, politics, and philosophy is replete with alternative guidelines and debates over the specification of standards for social justice in any given society. Nevertheless, there are both core themes in this literature and a significant contemporary consensus about the most basic elements of social justice, even as specific norms of social justice continue to be debated around the periphery.

If you are reading the chapters of the book sequentially, you will recall that we ended our discussion of sustainable development in chapter 5 by emphasizing the social ethic behind the Brundtland Commission's famous definition: development that meets the needs of the present generation while allowing future generations to meet their needs. There is a particular conception of social justice in that definition that appeals to fairness between people in present and future generations. In addition, this notion of social justice emphasizes needs over wants, preferences, or desires. These themes are common to many (but not all) ways in which justice has become a theme in studies of sustainability.

The Brundtland approach to social justice assumes a distinction between retributive and distributive justice, with responsibilities to future generations defined under the latter concept. *Retributive justice* concerns penalties or punishments that are exacted in response to moral infractions or violations of the law. *Distributive justice* concerns the manner in which goods and opportunities are dispersed within and throughout society. In stressing development that meets the needs of the present while allowing future generations to meet their needs, the Brundtland Commission interpreted sustainability as arriving at a fair distribution of earth's resources. The current generation cannot have it all, but at the same time, setting aside too much for future generations could pose an unfair burden on people living in the present. Recall from chapter 5 that global development aimed to redress the disparity between industrialized economies in Europe and North America and those of their former colonies in Africa, Latin America, and parts of Asia. Burdening countries that were at the early stages of industrial development doubled the distributive problem because it perpetuated an already unjust distribution of global wealth.

In general, there is a presumption that prospects for living well should be open to all and that, like the opening position in any game or sporting event, each participant has an equal chance at success. But as in games or sporting events, some types of inequality are not seen to create unfair advantages. One person or group who enjoys an edge that is not available to others may have earned it through prior struggle, work, or practice, or they may have unique talents or abilities that improve their chances. In general, an advantage is problematic when it derives from inequalities or favoritisms intrinsic to social rules, policies, or institutions. Policies or social practices that create structural or arbitrary disadvantages for certain groups are textbook examples of distributive injustice, but the details of distributive justice are hotly disputed.

At the same time that an advantaged person or group may have earned their access to better life chances, they may have inherited their privilege due to the theft, exploitation, or morally unjust actions of their ancestors. The generations who committed these injustices might be liable to fines or imprisonment if they were alive today. This suggests that they violated the requirements of retributive justice, but if they are no longer around to suffer the punishment, how is one to understand the situation? Groups subjected to violence and theft in the past can plausibly see their present plight as a continuing failure to punish offenders and offer compensation to victims. Given the role of colonialism, the slave trade, and European conquests in creating the disparities between the global North and the global South, it is possible to see the problem as a failure in retributive justice, even as the Brundtland approach interpreted it as a problem of just distribution. This tension between retributive and distributive orientations to the problem of justice adds a further complication to the meaning of sustainability.

How is social justice related to politics?

It depends on how one understands politics. One definition says that politics is activity that a group of people undertakes to form their sense of community. Politics enables joint or collective activity among people or the ability to act as "we," rather than "I." Joint action then binds people in a common fate. This notion of community and cooperative fate delimits a public sphere distinguished from types of action or conduct where members of the community are understood to pursue personal goals and private goods. Specifying a conception of social justice is central to this sense of politics. A conception of social justice specifies what people owe to one another in the public sphere and limits intrusion into the private sphere, where individuals are not constrained by the duties they have to other members of their political community.

When the word politics indicates a particular individual's or group's quest for power over others, social justice articulates important principles that the political community believes should limit that quest. In calling social justice an ideal, we are saying that it expresses or conceptualizes how the interaction of forming a political community should or ought to take place. Injustice occurs whenever actual practice falls short of this standard. However, in most contemporary communities, social justice is not simply equivalent to everything that people believe should or ought to take place. There are also norms and standards that apply to individual behavior or to groups (such as families) whose conduct lies outside the public sphere. It is more typical to reference such norms or standards as ethical or moral, rather than referring to them under the rubric of social justice, and we will follow that convention in this book.

There are other sub-forms of social justice beyond retributive and distributive justice. *Justice in recognition* is a subcategory that calls attention to the need for acknowledging differentiated forms of cultural, ethnic, gender, and religious identity. While for the most part justice requires treating everyone equally, there are group- or gender-specific needs that may need accommodation. The goal of equal treatment may not justify interference in the way subgroups maintain social cohesion and dignity. (For example, some religious communities enforce dietary restrictions or mandate certain types of clothing.) Policies that regulate a person's dress (such as requiring uniforms in public schools) might promote justice in a broad sense while compromising a subgroup's ability to maintain a unique identity. Another sub-class is *participatory justice*. Justice is not solely about access to goods or economic and cultural opportunities. It also requires that people have an opportunity to participate in decision-making that has a significant impact on their future, their culture, and their quality of life.

A recent surge of interest in problems of social justice parallels important aspects of sustainability. This began with a movement to address health disparities arising from disproportionate exposure to environmental pollution. Industrial production and waste disposal sites are often near poverty-stricken neighborhoods. These environmental health risks are problems in environmental justice. The idea of environmental justice has expanded and grown into a number of justice-based concepts that are appropriate topics for sustainability.

What is environmental justice?

The term *environmental justice* encompasses a large class of social problems that have origins in or are related to human impact on environmental quality. The idea couples the risks of exposure to pollution and resource depletion with the vulnerability that members of socially marginalized groups face in their inability to prevent or control exposure to these risks. Individuals from all social classes suffer from exposures to toxic substances and polluted water or air, and everyone is affected by depletion of natural resources. Suffering from the effects of pollution and resource depletion is a problem for environmental justice when being poor or racially identified causes someone's exposure or the harms they experience.

Understood narrowly, environmental justice refers to the policy and practice of government agencies, including the courts. Responding to an executive action by President Bill Clinton, US regulatory agencies work under a mandate to administer environmental laws fairly. This means that protections should not favor advantaged groups, such as white or wealthy Americans, and that remaining risks should not fall primarily on ethnic minorities or other traditionally disadvantaged groups. In addition, when injuries or public health impacts from environmental pollution are linked to discriminatory decision-making, anti-discrimination laws can be invoked to recover damages. These environmental justice provisions operate beyond the normal environmental regulations that protect all members of the public.

Understood broadly, environmental justice encompasses concerns that received comparatively little attention in public policy or in social activism prior to 1970. These include ways workplace hazards, access to food, and the distribution of environmental risks exhibit patterns similar to the way that oppressed people are unfairly deprived in more familiar domains, such as employment and educational opportunity, housing, or access to government services. In the United States, the social movement for environmental justice started in Afton, North Carolina, in 1982 when poor, rural, black residents organized to protest a hazardous waste facility being built there. As environmental justice problems have received greater attention, they have been associated with efforts to improve sustainability.

How is environmental justice part of sustainability?

Problems of environmental justice plague the permitting and construction of hazardous facilities. Refineries and factories that emit toxic substances into air or water are often located near populations of racial and ethnic minorities. The operators of such facilities may be motivated to reduce costs by locating a factory or waste disposal site where land values are already low. Impoverished neighborhoods inevitably have lower land values. The siting decision may also reflect strategic thinking on the part of operators: marginalized groups have limited ability to influence the permitting process. This simplifies obtaining approval to build a hazardous facility and reduces the chance that the operator will face a challenge in court. In the most egregious cases, public officials have interpreted regulations in ways that allow hazardous facilities to be located near black neighborhoods while preventing their construction in otherwise similar cases where risks would be borne by affluent white people.

If one takes the Brundtland approach to sustainability, the problem here is that environmental risks and health effects are unfairly distributed. Race and wealth should not be factors that cause one group of people to suffer more disease and injury from pollution or resource depletion than another group. Taking the perspective of retributive justice helps us see that racial minorities, along with women, suffer greater harm today because of the way they were subjected to abusive treatment in the past. Present-day injuries or disadvantages are then understood to be products of a system that perpetuates injustices that originated long ago. Distributive inequities in the present then come to be seen as resulting from past failures to observe the most basic norms of retributive justice: protecting the safety of everyone and punishing those who violate another person's property and threaten or perform acts of violence upon them.

What is food justice? How is it related to sustainability?

Food justice is a special class of environmental justice related to the production, distribution, and consumption of food. People who experience uncertainty in their access to adequate food, or even an inability to access food, are described as food insecure.

Food security is a problem in distributive justice, as poverty and reduced income are the most frequent causes of insecurity in food access. Because racial and ethnic minorities are disproportionately poor, food insecurity is also skewed along racial lines. In addition, gender biases within family or ethnic groups can create disparity in the degree to which women lack food security. Food security can also relate to the nutritional quality of available foods. While charitable organizations operate food pantries in many poor neighborhoods, these organizations may receive donations of unhealthy snacks and prepared foods in lieu of nutritious fresh fruits and vegetables.

In the United States, black, Native American, and Latino populations experience higher rates of diabetes and heart disease than the white population. These disproportionate rates of dietary disease reflect problems in food justice, though there are disagreements about the specific nature of the problem. One diagnosis emphasizes poverty. Grocery chains have often abandoned poor neighborhoods due to lower profit margins. This leaves the residents of such neighborhoods in a position of either traveling a significant distance to purchase food (often at significant expense in both time and money) or relying on less healthy convenience stores and fast food outlets for much of their daily diet. The lack of access to healthy, affordable food choices in certain parts of many urban areas has led some analysts and activists to describe these neighborhoods as food deserts (though people who live in these neighborhoods and people who live in real deserts object to this expression). What is more, food access is not a uniquely urban problem. In many regions, the rural poor must travel long distances to reach retail establishments that carry a variety of foods.

A retributive justice explanation of disproportionate rates of dietary disease stresses the way that colonizers oppressed victims by interfering in their ability to eat traditional ethnic diets. These practices operated through both material and cultural mechanisms. Enslaved Africans in the antebellum southern United States were forced to eat the leavings of the plantation system. They lost access to the diets of millet and greens that they would have eaten in Africa. Native American tribes were forced to eat low-quality foods such as fry bread that were supplied to reservations by government agents. As time wore on, unhealthy fried or sugared foods came to be associated with these ethnic identities. In the view of some activists, ethnic foods come to function as ways that historically marginalized groups participate in their own oppression.

In some cases, genetic analyses support this argument. Members of the Pima tribe in the southwest United States suffer from some of the highest rates of diabetes in the world. Studies support a genetic predisposition to this disease among the tribe. However, the Pima once cultivated the tepary bean, a variety with a very low glycemic index. They were prevented from growing these beans throughout most of the twentieth century. In their place, Pima cooks relied upon commodity bean varieties and processed foods for most of their dietary needs. Members of the tribe are again cultivating tepary beans and experimenting with the recovery of their traditional diets, and the rate of diabetes within the tribal community has decreased dramatically.

What is food sovereignty?

Food sovereignty refers to a cultural group's ability to have control over their food, often referring to the entire food system. Control must begin with primary production on farms and pastures or through harvesting wild plants, fish, and game. Control over the food system should extend through processing and distribution, right up through preparation and consumption. Some activists describe deprivation of access to culturally appropriate foods, as in the case of the Pima, as an affront to food sovereignty. In food policy discussions, food sovereignty is often asserted in opposition to proposals that define food justice solely as a problem of food security or access. Advocates of food sovereignty regard simply having access through purchase or through forms of food aid as insufficient. As a component of food justice, food sovereignty requires that communities can decide for themselves where, how, and with what they will meet their dietary needs.

Food sovereignty connects directly with environmental justice when a community or group is forced to abandon a culturally and nutritionally important source of food as the result of losing access to or control over the ecosystem that produced it. The Karuk tribe in Northern California has protested against management of former tribal territory by the US Forest Service, arguing that they have lost their ability to subsist on salmon fishing. Similarly, Anishinabek bands throughout the Great Lakes region have protested against the impact of white settlement on the marshes where they have traditionally harvested manoomin, or wild rice. In many of these cases, the injustice lies both in the loss of the ability to protect and maintain the resource and in efforts to undercut the community networks and knowledge systems that allowed tribal communities to lead healthy, productive lives centered around their practices of food production.

Beyond the United States, food sovereignty has been a rallying cry for Via Campesina, a confederation of groups aiming to protect small-scale farming communities around the globe, especially in Latin America. In this case, food sovereignty reflects a desire to maintain relationships between peasant farmers and the village residents who have traditionally eaten their crops. Food sovereignty opposes efforts to either consolidate these farms in pursuit of greater efficiencies or push for farmers to grow higher-value commodity crops (such as coffee, soybeans, or sugar) that must be sold on global commodity markets. These activists argue that such changes not only would weaken social and economic ties between farmers and community members but would also leave both parties at risk to both economic and nutritional harms visible among oppressed groups in the industrialized world.

What is climate justice? Why does it matter for sustainability?

Like food justice, climate justice refers to a specific class of
problems in social justice, this time related to the process of
global change. The stocks, flows, and feedbacks of energy and
greenhouse gases that underlie global climate change have already had significant warming effects at northern latitudes.
These changes are affecting the culture and way of life of
people who live above or near the Arctic Circle. The Sami way
of life has become more diverse, but it continues to revolve
around the reindeer. As precipitation in northern Scandinavia
shifts from snow to rain, freezing has covered pastures with
sheets of ice that disrupt grazing and threaten Sami husbandry.
Elsewhere, sea level rise threatens to disrupt subsistence production practices and, in extreme cases, can make populated
regions completely uninhabitable. The Marshall Islands are
a case in point. Analysts predict that much of the area where
native Marshallese currently reside will be underwater within
the lifetime of the current generation.

These climate change impacts are matters of distributive
justice because they disproportionately affect some groups
more than others. In many (but not all) cases, one group is
being harmed by climate change, while another is enjoying the
benefits of climate-forcing emissions. Those being harmed did
not derive the benefits from industrial emissions, either in the
form of direct utilization of manufactured goods and energy
use or through development that contributes to the overall
wealth of industrial societies. While everyone will suffer in
some way from climate change, most of us in the temperate
zones of industrial societies have benefitted from the economic
growth that accompanied the burning of fossil fuels. This is
therefore a straightforward case in which one group benefits at
the expense of another: a classic case of distributive injustice.

The impacts are problems for retributive justice because
the most severely affected people are not the people responsible for the emissions that cause climate change. From this

perspective, the harm one suffers from climate change is something that was done to you by someone else. That person should be punished, and they may need to offer reparations. If the emissions occurred in the distant past, there is no way for the legal system to punish them, but a retributive view might hold that victims of climate harms are owed reparations paid by those who benefited most from the unjust action of past generations. Of course, this is a controversial way to understand climate justice, but it is significant to note how taking a retributive perspective highlights systemic interaction in the way that benefits and harms are both produced and reproduced over time.

Many problems in climate justice are complex. Even the questions of whether and how the injustices just described should be resolved are debated. Do people who have already been harmed by warming or sea level rise have the right to seek remedies in international courts? Could they force emitters to cease? Do they have the right to claim damages for the harms they have suffered? Should they at least receive assistance from wealthy industrial nations in moving or making changes to their ways of life? Whether or not legal proceedings ever address these questions, they identify some of the ethical quandaries that accompany processes of global change.

Yet another form of climate injustice arises from the way that scientists and other climate activists have tried to motivate personal action or policy change to reduce climate-forcing emissions. The call to action on climate is almost always sounded as a plea to save the world from environmental catastrophe. This way of framing the ethical issues ignores the fact that for many poor and indigenous peoples around the world, the catastrophe has already occurred. They are currently involved in picking up the pieces and adapting to a world in which their traditional ways of life are no longer feasible. In presenting climate justice as the attempt to avoid some future calamity, climate activists fail to recognize the plight of people who are suffering today. This is a failure of justice in recognition that ignores and even conceals injustices currently being suffered by indigenous groups. It continues a pattern of marginalization that such groups have experienced since the early days of colonization.

How does social justice relate to sustainability?

The discussion so far has already highlighted quite a few different ways to interpret this question. The Brundtland Commission approached sustainable development as a forward-looking problem of distributing finite resources to future generations. Another view stresses past injustices committed by wealthy industrialized countries. If exploitation by wealthy countries caused the poverty of less industrialized nations, then sustainable development can be seen as a form of retributive justice: industrialized countries in some sense gained their wealth at the expense of the poor, so finding a path to compensate poor countries for these past wrongs is needed to rectify the situation. But given the view that future development is a function of the total capital at any given society's disposal, sustainable development can just as easily be interpreted as a problem in distributive justice. That is, setting aside for a moment the injustices of colonialism, slavery, and the exploitation of labor and natural resources, just giving people in less industrialized countries a fair chance at improving their standards of living becomes a problem of justice in a world where natural capital is limited (and probably declining).

These globally oriented ways of thinking do not exhaust the ways that sustainability can be linked to social justice. As we mentioned in chapter 1, policy change and social activism associated with sustainability encompass a very broad array of social goals. When groups lack access to basic needs or social services, the situation is decried as unsustainable. Community groups or local and state governments then undertake remedial efforts under the banner of improving sustainability. There is a bandwagon effect, where every social cause gets lumped together under sustainability. Yet civic or business leaders and activists alike view lack of access to goods and opportunities for racially and ethnically marginalized groups as unsustainable. This suggests something deeper than simply jumping on the sustainability bandwagon. Advocacy groups and local business councils or economic development organizations have all formulated action plans to address race- and gender-based inequalities under the rubric of sustainability.

In another time and place, similar activities might have been justified as promoting social justice (or rectifying injustice), so there is one sense in which sustainability is seen (by some) as encompassing or perhaps replacing social justice. In the United States, at least, this trend may be a response to the way that commentators on the political right have attacked efforts to redress past injustices or achieve a fairer distribution of resources and opportunities. Rather than challenging the way that activists interpret justice, personalities such as Glenn Beck and Rush Limbaugh have challenged the very idea of social justice itself, claiming that it is simply socialism. Perhaps in response, political leaders have chosen to pursue these social goals as components of sustainability rather than social justice.

Is social justice inherently tied to politically liberal or anticapitalist politics?

It is difficult to deny that today many people who identify themselves as liberal have a strong commitment to social justice, while people who identify as politically conservative are suspicious of appeals to social justice. However, when viewed historically and philosophically, the answer to this question is clearly, no. It is not that conservatives oppose social justice. Rather they have a different understanding of the role that governments should play in its pursuit.

The political philosophies that are today labeled as conservative have descended from ideas that were considered liberal in the history of European political thought. Conservatives are apprehensive about the use of planning and state power to achieve the personal liberties and opportunities for self-realization that are the goals of social justice. They see such actions as contrary to justice because they deprive individuals of their property and their ability to have control over their day-to-day activities. They also tend to believe that making voluntary trades is an expression of personal freedom; hence, they have an attraction to markets as a better mechanism of promoting justice than government programs. People who identify as liberal or anticapitalist have often found themselves disadvantaged by market forces. They have also seen how preservation of property rights has worked to the persistent disadvantage of women, ethnic minorities, and colonized people. As such, they are more open to public policies that remedy disparities and right historical wrongs.

In short, the division between conservative and liberal politics does indeed reflect different conceptions of social justice. This means that conservatives might favor charity, voluntary action, and markets as the preferred response to some of the social justice problems discussed in this chapter, while liberals will be more willing to try legislative solutions and public action. In no case does it imply that social justice is not a component of sustainability, much less that social justice is not a problem at all.

But there is more. The idea that a continuous stream of abuses threatens the stability or peacefulness of any society has a long history in both conservative and liberal political traditions. When people are hungry, deprived, or oppressed, they eventually get to the point where they have had enough, and periods of protest and rebellion occur. These outbursts can put more strain on social services (such as police and fire departments), and they disrupt normal activity. Thus, in a sense, injustice is a driver of crime, violence, and even revolutionary activity in society. When the outrage of oppressed groups reaches a boiling point, the institutions of a society can be driven into a state of collapse. Here, injustice threatens the system of social institutions that are thought to produce social sustainability. This approach treats justice (or injustice) as something rather like a stock that is affected by feedback from other activities and processes at work in the social system.

What is social sustainability?

As with the previous question, there is ambiguity in the way that social sustainability is interpreted. Wikipedia, a venerable source for wisdom on everything, states in the article on the topic that social sustainability is "the least defined and least understood of the different ways of approaching sustainability and sustainable development." From the perspective of business management, social sustainability is usually defined in terms of ideas like social capital or permission to operate that were discussed in chapter 2. From an environmentalist perspective, social sustainability covers the human impact associated with disruption of ecosystems and declines in environmental quality. Here, social sustainability is a category that links environmental disruption to standard concepts of social justice, including environmental, food, and climate justice.

Another interpretation takes the idea of sustainability more deeply into the formation and reproduction of social institutions, norms, and cultures. The processes that support social interaction depend on stocks, flows, and feedbacks of information, cooperation, and governance activity. These processes can be evaluated as more or less sustainable in themselves. Social sustainability is then a measure of whether and how long anyone can expect the important institutions, values, and norms of their society to continue. A period of normalcy in social relations would indicate sustainability. Violence in the form of crime or random acts of destruction is a stock kept at low and controllable levels. When the overall level of this stock is kept low, people are not afraid to go about their daily business in a routine manner. However one understands routine or normal social activity (and this will itself vary from place to place), one can think of it as something that can continue from day to day, and alternatively as something that can be disrupted by demonstrations, acts of resistance, and destruction of property. If these disruptive activities become extreme (that is, if the stock of disruptive acts reaches a tipping point), society itself can enter a state of collapse.

At the same time, both scholars and activists who have promoted the idea of social sustainability have often used the term to identify social changes that they endorse, even when there is no obvious connection to the environment or any threat of disruption or collapse in social routines. In such contexts, the term *social sustainability* might be applied to virtually any change in policy or practice that can be recommended on ethical grounds. A cynic might say that sustainability has just become a catch-all term for any and all forms of progressive social change. A more positive way to say roughly the same thing is that as people become more and more interested in sustainability, all manner of social activism becomes associated with it. Issues about fairness or oppression that have been swept under the rug for too long start to enter the discussion. There is a feedback loop—a virtuous rather than a vicious circle—that continuously reinforces public cognizance of past injustices and feeds the will to disrupt flows that reproduce injustice from one generation to the next.

Examples are likely to be contentious and even offensive to some readers, yet it is important to consider at least a few possible examples in more detail. One case in point would be structural racism, which we will understand to include systemic processes that continuously reproduce racial inequalities. Some of the more pernicious aspects of structural racism do this even after people take measures to combat racial bias. This makes inequality seem like a normal and natural course of events. US real estate markets are an oft-cited example, as race-based policies that were commonplace in the 1940s and 1950s led to highly segregated neighborhoods in many American cities. Homes in white neighborhoods received a boost in value that has continued to multiply for decades after these policies were outlawed. Since home ownership has been a dominant source of wealth for middle-class Americans, this cycle reproduces race-based social inequalities, even in the absence of racist intent. The question in this context is: How does structural racism figure in accounts of social sustainability?

A fairly common answer to this question is that social sustainability requires the dismantling of structural racism. Reparations for affronts of past decades might be referenced as one policy response, or new efforts to rebuild the infrastructure in still-segregated neighborhoods might be another. Since racial segregation is a factor in unequal resources dedicated to education and healthcare in the United States, improvements in the welfare of the black population are plausibly dependent on some form of change in policy and practice that will break the cycle of poverty, ending unequal access to resources. Inequality is a basic problem for distributive justice. This is not to say that society's wealth should be parceled out equally, but distributions should not be skewed by arbitrary factors. Historically, race is one of those arbitrary factors. If sustainability is defined as a form of sustainable development, it is plausible to think that dismantling the structures that reproduce race-based inequalities will contribute to economic and personal development among oppressed groups, redressing the unfair distribution caused by oppression in the past. If opposing structural racism is a social practice essential for sustaining development, wouldn't it also be part of social sustainability?

Yet one may question how sustainability is being understood in this way of thinking. There are already powerful reasons for addressing persistent inequalities, and especially so when they can be traced to patently unjust forms of racial prejudice that existed in the past. How does the idea of sustainability add additional power or persuasiveness to these reasons? One answer goes back to the theme of stability: if injustices are not redressed, they will simmer until violence breaks out and some form of rebellion ensues. A second answer is that when society decides to use the word sustainability as a blanket term for all good and noble things, it is natural that redressing the harms of structural racism will fall under it. Both of these answers are at work in the thought that social sustainability contains or implies some form of action against structural racism.

In fact, it is impolitic to argue against the idea that social sustainability implies action to reverse or correct the effects of structural racism. Any statement to that effect will very likely be interpreted as part of the normalization process that allows structural racism to endure. Yet the systems thinking approach behind sustainability does present another alternative. One could ask: What are the elements that have made the reproduction of these inequalities so robust and resilient, even in the face of actions that were intended to end them? This question suggests that the problem here is not so much that structural racism is unsustainable. The problem is that it is all too sustainable.

Posing the problem in this way takes us back to a question from chapter 1: Is sustainability always a good thing? In the domain of social justice and social sustainability, there has been a persistent tendency to presume that it is. While anyone (and that includes your authors) would be well advised to tread lightly when discussing the sustainability of social evils, the theoretical and analytical power of sustainability as a way of thinking may be sacrificed when people simply assume that sustainability is a synonym for "good" or "just." With this cautionary note behind us, we can pursue a series of questions that arise in trying to grasp the relationship between sustainability and social justice.

Can sustainability combat social injustice?

Yes. Sustainability gives us the ideas for analyzing systems that reproduce unjust distributions of resources, failures of recognition, and denial of opportunity to participate in social life. Using stocks, flows, feedbacks, and hierarchy, we can begin to paint a picture of why unjust circumstances persist from one generation to the next and why patterns of inequality and oppression are so resilient. Leverage points in these systems direct us toward more promising places to intervene in disrupting them. These concepts are not a be-all and end-all recipe for social justice, but they can help to promote a wider appreciation of where to attack the institutions that lead to unjust outcomes.

As already noted, the role of home values in structural racism provides an example. As early as the 1930s, banks and government lenders adopted the practice of "redlining." They literally drew red lines on maps to indicate neighborhoods where loans either would be subject to higher interest rates or would not be made at all. After World War II, the practice solidified the segregation of races in American cities, such as Detroit and Chicago, condemning blacks and other racial minorities to low-value homes or to renting properties owned by absentee landlords. Banks also denied them access to loans that would have allowed them to make improvements and necessary repairs on their properties.[1] These injustices were bad, but it is the systemic role that real estate plays for middle-class accumulation of wealth in America and other Western societies that constitutes structural injustice.

Growth in home values has contributed to increasing wealth for members of the predominantly white middle class. While individuals of all races make fortunes and gain wealth through talent, savings, and hard work, for most Americans living in the last half of the twentieth century, the ability to make payments on a home increasing in value allowed them to transfer significant wealth to their children. Homeowners who purchased a house for $20,000 in 1960 would repay that amount through set payments (perhaps $200 per month), but by the time the loan was paid off in 1980, the $200 monthly payment was less than what it had been in 1960 due to inflationary feedbacks from the overall economy. This is another example of hierarchy (the real estate market is a subsystem of the larger economic system). Homeowners effectively got a bonus because of the way that a higher-order system affected a subsystem in the hierarchy. By 1990 that same house would have been worth $60,000 to $200,000, depending on location. Homeowners' wealth was increased substantially, while renters did not enjoy the benefits of these feedbacks. This allowed homeowners (who were disproportionately white) to transfer enough wealth to children so that they too could make the down payment needed to enter this amazing system of monetary stocks and flows.

But of course that is not the end of the story. Real estate values are affected by feedback from other stocks, including tax revenues that allow municipalities to maintain infrastructure and make improvements. Home values in turn affect school funding, so that districts with lower home values also have a lower base for school buildings, supplies, and teacher salaries. They affect an individual's access to credit for both purchases that improve quality of life and loans that might be used to start a business or finance a college education. The total system of flows that is the modern metropolitan economy is dramatically affected by feedbacks that emanate from home value. People who were frozen out of home ownership or limited to low-value homes at the early stages of this systemic process were also impacted by the continuing feedbacks that affected their access to a host of goods essential to personal well-being in the cities of the twenty-first century. It is in this sense that structural racism is all too sustainable.

This sustainability analysis has two features that can combat injustice. First, it demonstrates how social institutions that produce racially structured inequalities are not a function of indolence, lack of talent, or some character flaw in the population that is on the short end of these feedbacks. It should undercut any feeling that people are getting what they deserve. It shows how racial injustice is a resilient feature of the system, even after measures have been taken to undo the policies and practices that initiated the injustice in the first place. Whites who feel no animosity toward blacks should not be complacent or believe that racism does not continue to produce inequalities. Simply making it clear that structural features of the economy and system of public finance reproduce a significant component of our racial inequality should open minds toward opportunities for making racial injustice less resilient.

Second, it suggests that the leverage points for change may lie in the nexus of feedbacks that radiate from the core stock of home values. It suggests that continuing to structure school districts and school funding around taxes on residential real estate is not a just practice, even if it appears to be something that can be continued indefinitely (i.e., is all too sustainable). At the same time, when real estate values determine local budgets, communities with low housing prices will have significantly reduced opportunities for undertaking change. This is not to say that making changes in these systems will be politically easy, but it does point us toward places where change could make a significant difference. The case of real estate and structural racism is just one instance where the idea toolkit of sustainability can make a difference for social justice.

Are there conflicts or contradictions between sustainability and social justice?

The question reflects some of the points already discussed, albeit with a slightly different emphasis. In the 1960s, ecologists predicted that human population growth would exceed the carrying capacity of the global ecosystem. In short, the rate of increase in the human population was judged to be unsustainable. This prompted coercive efforts to curtail the birth rate, most famously China's one child policy and recommendations that hungry people should be allowed to starve. But many argued that government population control violated human rights, and these policies had unanticipated consequences. In retrospect these actions intended to promote sustainability were judged to be unjust.

The Brundtland Commission was formed to address the contradiction between continuous global economic development needed to support growing populations and the recognition that pollution and the depletion of resources were placing limits on development. It started from the premise that the unequal levels of economic development between Europe, North America, and other industrialized countries on the one hand and the former colonies of South America, India, and Africa as well as less industrialized regions in Asia on the other was unjust. To repeat points already made, it was unjust both because it represented an unequal distribution of goods and because people in less industrialized regions had endured many years of coercive treatment. Concisely, the Brundtland Commission was charged with addressing the problem of social injustice on an international scale.

Their solution was to endorse development that meets the needs of current generations while allowing future generations to meet their needs. Or to put it slightly differently, sustainable development does not permit forms of economic development that would compromise future generations' ability to have a decent quality of life. Hidden beneath these simple statements is the assumption that economic development is the most promising route to resolving the social injustices that the Brundtland Commission was created to rectify. But since economic development could cause injustice to people living in the future, sustainable development did indeed imply that at least some of the routes to an increase in GDP were unacceptable. In this sense, there is clearly a contradiction between sustainability and some of the strategies that could be followed to promote social justice in international affairs.

The pattern of thinking that followed the Brundtland report has led many to presume that sustainability and social justice are not fully compatible. At the same time, the questions already covered in this chapter suggest that other people were starting to define sustainability in terms of policy or practice that promotes social justice. In addition, the thought that social injustice can spark revolutions or violence provides yet another reason to presume that a fully comprehensive notion of sustainability would have to include tenets of social justice. These ways of understanding what is required by justice seem to bring it more fully in line with sustainability. Yet our suggestion that injustice can be all too sustainable (when viewed in systems terms) implies yet another way in which sustainability and social justice might be understood as opposing ideas.

There is also an important sense in which systemically organized stocks, flows, and feedbacks incorporate contradictory tendencies to create systems of processes or practices that are robust, resilient, and adaptive. Consider predator-prey relationships as an example. The interests of predator species and the interests of the plants and animals that they consume for food are clearly at odds with one another. Yet in a balanced ecosystem predation controls the population stock of prey species in ways that allow the entire ecosystem to flourish. This is a deeper, more philosophical, sense in which someone might say that certain types of contradictions are inherent in sustainable systems. Would this also imply that social injustices must simply be accepted as the price one must pay to achieve sustainability?

We think not, though we admit that there are very hard questions here that we are not able to answer in a fully satisfactory manner. When we described social justice as an ideal at the beginning of this chapter, we recognized that there would be cases where it was not fully achieved. Yet precisely because it *is* an ideal, no one should give up on the attempt to redress injustices when they occur. Everyone should try to reconcile social justice with sustainability at both theoretical and practical levels. When practical steps toward sustainability are discovered to cause injustice, it is thus important to find ways to revise one's strategy and to practice a restorative justice that compensates those who have been unjustly served. Systemic interactions being what they are, one should not be surprised that specific attempts to promote sustainability will fail. Sometimes they will impede the pursuit of a just society. But this does not imply an irresolvable contradiction between these two social ideals.

Does environmental sustainability require a commitment to social justice?

As with the answers to most of the other questions in this chapter, people disagree. There are at least two ways of answering the question in the affirmative. First, poverty and inequality limit people's options. People who see the environmental threats of unsustainable lifestyles as the dominant issue can still admit that it will be difficult (if not impossible) to make the behavioral changes that are needed to be more sustainable without some effort to remediate problems of social injustice. Population growth provides an example. As previously discussed, unconstrained growth of the human population has long been recognized as a threat to sustainability. Demographers have shown that women's literacy is one of the key leverage points for lower birthrates. As women become educated and attain a greater degree of social and economic independence, birthrates decline. Here, addressing a problem in gender-based social justice is a tool for controlling population growth, thereby improving environmental sustainability.

Alternatively, many advocates of social sustainability (as well as advocates of "three circle" sustainability; see chapter 1) would simply assert that improvements in social justice are essential. They would object to the suggestion that environmental dimensions of sustainability have any kind of priority over longstanding goals of promoting social justice. This view of the connections between environmental quality and social justice is one basis for opposing the trend to substitute resilience for sustainability. The turn to resilience looks suspiciously like a return to scientific positivism, where scientists simply refused to acknowledge the relevance of value judgments in their research. In fact, critics of the resilience turn suggest that it hides a value judgment to the effect that social justice goals (and the interests of poor people or marginalized groups) can continue to be ignored. Wealthy people or privileged groups would then be justified in adopting policies to increase ecological resilience, even when doing so occurs at the expense of the less well off.

This criticism already suggests one way in which someone might answer the question negatively. Ecologists and environmental scientists have identified stocks, flows, and feedbacks that are crucial to ecosystem sustainability and environmental quality. They have not needed to undertake investigations of social justice to do so. Although many of these scientists would agree that social justice deserves a high priority in the pursuit of sustainability, they believe that environmental and social sustainability are logically distinct. Their support of social justice is based strictly on ethical grounds and has nothing to do with the science of sustainable ecosystems. And as already indicated, there are political pundits who have set themselves up as opponents of sustainability. Environmental scientists may be hoping that by dissociating environmental goals from social justice, they can appeal to a broader spectrum of public opinion.

Does sustainability require that people come to an agreement on their understanding of social justice?

This chapter has moved progressively toward more general questions of political philosophy, and there is a danger of losing our focus on sustainability in the process. This question points toward a debate that extends well beyond the bounds of a book on sustainability. The answer that has been favored by both liberal and conservative philosophers who maintain a commitment to either political democracy or scientific inquiry has been that tolerance for differing points of view improves the outcome of political debate, just as it helps a scientific community converge on the best explanation for the phenomena they study. If we adapt this thought to the present context, disagreements make the processes and practices people hope to continue more sustainable because the very process of disagreement becomes a way of testing ideas and promoting communal action. This presumes a willingness to listen to the ideas of others, of course, as well as openness to changing one's views when presented with persuasive arguments.

The opposing view has been held by despots and elites attempting to maintain their authority in public decision-making by stifling all opposing points of view. This is not the place to join this larger philosophical debate, except to note that the matter of tolerating different views on social justice has been at the heart of it, at least since Plato. There is, in conclusion, a long history of thought holding that our society will be more sustainable if it has the capacity to maintain a lively debate over the terms of social justice and the particular strategies for promoting it in any given situation. The premise that fostering debate about social goals and implementing strategies for sustainability requires a sustainable system of governance is the topic of the next chapter.

7

SUSTAINABLE GOVERNANCE

Why do we have to talk about government?

This is a chapter about government. At all levels, governments are charged with providing a broad range of public services to meet the demands of a diverse citizenry. The ability of governments at multiple levels to meet these challenges and responsibilities represents a problem of sustainability. Every book about sustainability includes an obligatory chapter about what government needs to do achieve sustainability goals. That's not what this chapter is about. If governments are to meet their responsibilities of helping societies meet basic needs and improve the quality of life, then governments must be sustainable. If governments are badly run or badly organized, the services they provide will not continue. Sustainability of government is thus a measure of whether and to what extent services critical to basic needs and quality of life can continue.

Of course, we can't avoid talking about the role that government policies play in meeting economic or environmental sustainability goals, and following from our discussion in chapter 6, government actions are deeply interwoven with social justice. We also recognize that some readers expect the whole book to be about what governments should do. If we haven't disappointed those readers so thoroughly that they gave up on the book several chapters ago, it would be a shame to lose them here. In chapter 6, we discussed how different views on the role and capability of government action undergird competing conceptions of social justice. That discussion carries over to this chapter. History supplies many cases where governments adopted policies to address social concerns, but the policies triggered feedback that exacerbated the problem. Prohibition in the United States during the 1920s was aimed at reducing social problems associated with alcohol consumption; instead, bootlegging, smuggling, and organized crime proliferated, and speakeasies were the place to be.

What do governments do to increase sustainability?

In previous chapters, we discussed sustainability indicators for environmental quality and economic development. Governments can take measures intended to move these indicators. First, they can ban or restrict flows (like noxious emissions or use of chemicals). Second, they can administer programs intended to increase stocks (such as healthcare or food access). Third, they can use taxes, fees, and interest rates to incentivize behavior that positively influences indicators for environmental quality, economic development, or social justice. Fourth, governments can create tradeable rights to discharge wastes into the air or water or use resources (such as water) that allow profit seekers to bargain with one another. This strategy can create sustainability incentives while avoiding the coercive nature of regulatory mandates, and it helps to allocate scarce resources to leverage points where the impact on sustainability will be greatest. The universe of specific policy proposals to promote sustainability is large. Debating and then adopting policies is, indeed, important, but placing government into a systems context will do more.

Governments execute the adoption and enforcement of policies in a social environment where people push governments toward the policies they favor. Some people are motivated by personal or pecuniary interests; others choose which proposal to support based on ideology or philosophical vision. The mix of interests attempting to influence policy can become quite convoluted. For example, the Obama administration issued new fuel efficiency standards for passenger cars as a component of environmental policy intended to reduce greenhouse gas emissions. The automobile industry petitioned the Trump administration to relax these standards or delay the deadline for implementation. In keeping with President Trump's views on regulation, the decision was to roll back the new requirements entirely in 2020. At this point, several auto manufacturers decided to join forces with those who were opposing this action from the Trump administration. The auto companies felt that a total lack of regulation would deprive them of the ability to plan for changes in their automotive designs. This example illustrates how economic interests interact with political ideology, especially when there is uncertainty as to how any given policy will perform. Political influences mimic the stocks, flows, and feedbacks that we have discussed in other chapters. That is why it is important to look beyond policies that are intended to move indicators in the direction suggested by sustainability and to consider the larger question of sustainable governance.

In addition, the question of what governments can do depends mightily on what type of government you are talking about. For example, a president, prime minister, monarch, or party head in less democratic states can take actions to promote environmental quality without fearing that their environmental policies will undermine their ability to perform other functions of government. Policies to reduce pollution and resource depletion may not be politically sustainable in democracies, because a regime that puts such policies in place will face so much outrage that they will be voted out of office or the action will be overturned through legal challenges. Yet we aren't saying that democracy is unsustainable, or even that democracies fare worse on environmental policies than an autocratic state. Depletion of the Aral Sea (discussed in chapter 4) was government policy in the old Soviet Union. There are many similar counterexamples to the thought that less democracy is good for sustainability. The point here is that one cannot recommend any policy without taking the entire system of governance into account. We are embedding questions about what governments can do in a larger discussion of sustainable governance. The language of systems has permeated the chapters of this book so far, and government is but one part of systems of governance. This chapter is about the sustainability of governance systems.

What is governance?

You won't find one simple definition of the term *governance*, but almost all of them emphasize a process of deciding on collective goals and designing the means by which those goals will be met. Our objective in this chapter is to explore how governance processes do and don't work and what that means for the ability of governments to satisfactorily do those things that citizens believe they should do. Each of us is familiar with governance in some form. Those of us who are members of a civic organization know that the organization has a constitution or bylaws agreed to by the members. Such documents lay out the purpose of the organization and describe how that purpose will be pursued. They usually include procedures for a governing body (elected officers) that oversees the running of the organization and its activities. If you work for a company with a president, vice president, and a group of division managers and assistant managers, you are familiar with governance in that context. In this case, a formal process and structure exist that prescribe the objectives for the company's day-to-day operations and its future and specify who is responsible for overseeing efforts to meet those objectives. Colleges and universities have structures for academic governance. Finally, while most of us don't think about it too closely, our towns, cities, states, and countries function under systems of governance. Governance is not just government by another name. Instead, a government is part of a larger system of governance, just as the elected officers of an organization or the directors and managers of a corporation are part of a larger system of organizational or corporate governance.

Many of our decisions have positive or negative effects on someone else who had no part in the decision. If I plant an oak tree, my neighbor can enjoy the shade; if my neighbor decides to put up a beehive, I might be the one who gets stung. Governance can be understood as the process for negotiating benefits people enjoy and risks people bear as a result of each other's decisions. In many cases, these negotiations can take place without involving government at all. In fact, complex societies have developed nongovernmental institutions that range from garden clubs and homeowners' associations to trade unions and the Better Business Bureau. All of them perform functions that can be understood as governance. Governments get involved if there is no other way to resolve the problem when someone who is providing a benefit to others wants to get paid for it, or when some action imposes a cost for which someone wants to be compensated. Unfortunately, people have different views on whether those being affected by the decisions can get together and work it out on their own. The problem with pollution, for example, is that its effects are often widespread, and it gets hard to have successful negotiation between polluters and those affected. And the negotiating power of those being affected is diluted as their number grows. That is why governance decisions about such things generally vest authority in governmental units.

Boating provides a metaphor for the distinction between governance and government. Governance is the steering (or purposing) process that guides the rowing (or performing) actions of government. In the context of garden clubs or homeowners' associations, there may not be a big difference between rowing and steering. In the context of a local, regional, or national government, governance is the process by which members of a community or society work together to identify social goals, solve social problems, and design and manage the infrastructure (usually government) created to do the work toward achieving goals and solutions that individuals cannot do on their own—like resolve pollution problems. Somewhat more abstractly, governance processes require choices about things like (a) where power in the governance process is vested (who gets to help steer), (b) what is needed in order to improve society in ways that individuals alone cannot do (settling on and steering toward the destination), (c) who is charged with doing the work needed to pursue the goals (who gets to help row), and (d) how that work will be resourced and guided (who pays for the boat, the paddles, and the rowers and makes sure they are maintained in good working order).

What is sustainable governance?

In short, sustainable governance is a governance system that is able to continue. Machiavelli's sixteenth-century book *The Prince* was a handbook for leaders to follow if they want to maintain their power. His advice was that it is better to be feared than loved. However, in common usage, the term *sustainable governance* carries with it a connotation of good governance, or at least governance that is steering a government and monitoring its activities in a way that is acceptable to those being governed. Unarguably, there are systems of governance—and governments—around the globe that are startlingly sustainable (that is, they have been in place for a very long time) despite clear evidence that the majority of the population being governed would fare better under another system. Machiavelli might feel vindicated if he were alive today. Thus, to be more precise, sustainable governance is able to continue indefinitely to the benefit of the society being governed. Not only is the governance system sustained, but so too are the government and its activities that satisfy the needs and wants of those being governed. (Of course, this is a generalization. Satisfying everyone is rare.)

Monarchies and non-democratic states aside, we believe that collaboration and inclusive decision-making are an integral part of sustainable governance. A system of governance in which private citizens play particularly prominent roles in both the steering tasks of the governance system and the rowing tasks of formal governments is more likely to continue indefinitely. A system in which public and private sectors collaborate in the public interest can meet the demands of those being governed more effectively. Collaboration among the business sector, nonprofit organizations, individual citizens, and formal government offers the opportunity for more innovation in governance structures and government activities. The availability of multiple actors to complement and backstop each other introduces robustness and resilience in the face of changing economic and political conditions and changing social priorities. Because formal governments are facing political polarization and global uncertainties and working with limited financial resources, the sustainability of public initiatives in the interest of achieving social goals is enhanced when responsibilities are vested across a broader set of actors than just the elected and appointed individuals traditionally charged with governing. Using our systems language, collaboration among public and private actors is the source of inflows into stocks of legitimacy, financial resources, and knowledge to support actions taken toward public goals.

The sustainability of a governance system isn't an all or nothing proposition. There are countless examples of public policies, programs, and initiatives that have proved unsustainable even as the larger governance system has continued to function. Where these policies and programs have addressed issues of critical concern to the citizens affected, their failure—and the failure to achieve the social goals they target—raises smaller-scale questions of sustainability. As such, the sustainability, or lack thereof, of governance systems is reinforced at multiple scales (an example of hierarchy, a concept we discussed previously), from the robustness and resilience of the full system of governance to the effectiveness of individual public actions taken in pursuit of public benefits. Individuals making decisions about how to vote or contribute money to a political cause operate at one level of the hierarchy. They may not have much appreciation of the way their choices reverberate at other levels. We believe that governments could do more to promote sustainability if more individuals could view their political behavior from a systems perspective.

Are there indicators of sustainable governance?

Yes. In previous chapters we've talked about environmental indicators and economic indicators; this adds to that list. Bertelsmann Stiftung, a private German foundation, works with a panel of experts to analyze and report on governance systems in the forty-one OECD countries across the world using a set of sustainable governance indicators (SGI).[1] They use data from a cross-national survey to generate a set of qualitative and quantitative indicators of sustainable governance across three main areas: policy performance, democracy, and governance.

The policy performance index assesses policy initiatives and actions across economic, social, and environmental policy priorities. This set of indicators aligns with some of the economic, environmental, and social justice issues discussed previously. The index reflects the view that whether and how governance systems successfully pursue social goals related to environmental health and human welfare (sustainability goals) are important elements of their own sustainability. When such goals are clearly articulated but ignored, governmental actions are called into question. How do some of the forty-one countries compare in terms of their indicator scores in this area? The United States does not fare well in the evaluation of economic policies (ranked twenty-fourth), social policies (twenty-ninth), or environmental policies (ranked at the bottom). Federal budgetary policy receives a low score, 3.8 out of 10, due primarily to the budget deficit. Child poverty is a social policy challenge in the United States. Failure to lead, or even participate in, global climate change mitigation efforts drives the poor showing on environmental policy. In contrast, Sweden ranks at the top for economic policies (scoring 8.1 out of 10), with strong budgetary policy and strong research and development policy. Norway ranks at the top for social policies because of its wide range of social programs: poverty is low and educational attainment is high. Sweden ranks at the top for environmental policy as well (scoring 8.7 out of 10).

The democracy index evaluates the robustness of democratic institutions and practices, and the governance index examines the steering capabilities of governance structures and the extent to which nongovernmental actors are involved in policymaking. Indicators underlying the democracy index assess electoral processes, protection of civil and political liberties, integrity of legal institutions, and citizen access to government information. Scandinavian countries receive the highest scores for quality of democracy; the United States ranks fifteenth. The United Kingdom ranks eighteenth. Hungary and Turkey rank fortieth and forty-first, respectively, for the democracy index. The governance indicators explore the capacity of citizens for participation in governance and their degree of involvement. They also assess resources available for effective governance, capacity of nongovernment groups to participate in policymaking, and thoroughness of media coverage of government actions. Scandinavian countries also rank at the top for the governance indicators. The United States ranks relatively low (twenty-eighth) for executive capacity, or preparedness of the federal government to meet planning and policy challenges, but ranks seventh for effective involvement of nongovernmental actors in policymaking.

The SGI project focuses only on national governments. Absent a similar synthesis of indicators for cities and states, one can explore a host of online "best and worst" lists compiled by public and private entities. These lists evaluate cities and states according to things like economic health and fiscal stability, public health and safety, quality of public services, operating efficiency, and government transparency and accountability. We include nongovernmental associations in the list of institutional actors that could be evaluated. Groups that work to influence government decision-making (trade organizations and social advocacy coalitions, for example) and organizations that perform steering and rowing functions (standard-setting bodies, discussed below, or certifiers such as Underwriters Laboratories (UL), which tests electrical appliances for safety). Although there are indicators for sustainable governance, there is also more work to do in this area.

What are standards, and how are they part of collaborative governance?

Answering this question will take a bit of effort. Here are a few quick examples of standards: If you go to the deli counter at your local supermarket and order a pound (or kilogram) of sliced smoked turkey breast, you know how much you are getting because of the standardization of what a pound (or kilogram) means. If you purchase a hairdryer anywhere in North America, you can use it anywhere in North America. But if you travel between North America and Europe, you need to take along an adaptor because the electrical sockets look different, and you may even need a converter because the standard voltage for electricity in European countries is 220 volts while it is 110–120 volts in North American countries, and not all hairdryers have converters built into their electrical wiring. Cell phones are similar; two competing standards (Apple and Android) require special programming to interact. Fortunately, individuals don't have to worry about this compatibility problem, but it is a big problem for companies that develop applications. (You should begin to see why standards are important.)

Some standards are developed by governments, but others are developed privately. The International Organization for Standardization (ISO) oversees the development of standards for all types of industrial activities. In fact, businesses of all types, as well as government agencies, universities, and other organizations, also use the standards established by ISO. The ISO industrial standards address everything from units of measure to methods of testing equipment and materials, from manufacturing processes and packaging protocols to transportation infrastructure and waste management. ISO standards are written and reviewed by experts from business and industry, government, and academia, but ISO itself is not a government entity. ISO supports its activities by charging a licensing fee to anyone who wants to use one of their standards (and tell customers or regulatory bodies that they do).

ISO 14001 is the standard developed for environmental management systems that companies (and government agencies and other entities) use to minimize the environmental impacts of all facets of their operations: air emissions, wastewater discharges, solid wastes, toxic wastes, and so on. This standard provides a good example of how businesses are involved in environmental governance. Many environmental regulatory agencies provide businesses certified as meeting the ISO 14001 standard with expedited permitting processes, less extensive reporting requirements, and more streamlined inspections because the certification provides assurances about how the businesses are operating. This also reduces the burden on environmental regulators, since they can focus more of their efforts on operations that cannot demonstrate the level of environmental protection that certified businesses can.

Standards are also developed as part of a private firm's business activity. Apple's iPhone provides an example. The iPhone is, in fact, a pocket computer that can do many things. In order to make telephone calls, it must comply with ISDN standards, established in 1988 by an agency of the United Nations. In the age of the Internet, ISDN is a component of an even larger Internet protocol suite known as TCP/IP. Readers may have seen these acronyms without wondering what they stand for, and even superficial discussions of the standards needed to make computers interact with one another (plus video streaming services, television sets, and other devices) become mindboggling. iPhone OS (the program that makes this pocket computer work) was developed to comply with these previously existing standards for electronic signals, but it in turn became a standard for other companies developing software applications (games, utilities, and communication services). Apple chose not to license iOS to other cell phone makers, giving Google the opportunity to develop Android. If you want to build, use, or develop new applications for cell phones today, you must comply with one of these two standards.

Although the process for developing and implementing standards is largely invisible to the average person, one *must* use tools and products that comply with standards in order to live in today's world. As noted, some standards are developed and enforced by government. The US Organic Standard was developed by the Department of Agriculture, for example. In other cases, governments mandate the use of standards developed by nongovernmental organizations. The National Fire Protection Association is a nongovernmental organization that develops and tests standards for fire safety, yet state and local building codes in the United States generally require compliance with NFPA standards. Standards are thus a part of governance, without being a part of government. They are sometimes referred to as soft law—a system that structures our behavior but does not have the coercive authority of the police to do so. Soft law and standards play a very large role in enforcing compliance with sustainability indicators.

How do standards promote governance for sustainability?

We have already discussed ISO 14001, which identifies environmental quality indicators that track what certified managers are doing in their day-to-day decision-making. Other nongovernmental organizations develop standards and conduct certifications across a range of activities. LEED standards (Leadership in Energy and Environmental Design), developed by the Green Building Council, are among the best known. LEED standards include a list of criteria that builders can fulfill to accumulate points toward increasingly more stringent certifications (e.g., silver, gold, platinum). Firms, individuals, and organizations construct or renovate facilities that comply with these criteria, then pay a fee to the Green Building Council to receive their certification. Their motivation is partly altruistic—they want to do the right thing for sustainability—but firms and organizations like hospitals, schools, and governments also do it to enhance their reputation. They want others to see them as practicing environmental citizenship.

Readers will surely have some familiarity with Fair Trade, a label attached to many consumer products, especially coffee. The label certifies that the product was made in compliance with workplace standards and fair returns to small farmers. Other standards are emerging in the food system. Food Alliance is a nonprofit organization in the United States that represents a collaboration of universities, state governments, foundations, and private industry to advance sustainable agriculture. It began its operations certifying farms that meet standards for sustainable agricultural production practices. Over time, Food Alliance expanded to certify agricultural producers and food companies that follow standards for fair labor practices, humane treatment of animals, wildlife habitat protection, and food processing and distribution, among others. Certification through Food Alliance provides consumers with information about how food is produced by participating farms and companies, information that would not be readily available otherwise.

The Fair Labor Association (FLA) is an international non-governmental organization that works through a collaboration of universities, nonprofits, and companies to address labor issues in factories and on farms around the world. FLA helps participating companies implement a workplace code of conduct that promotes compliance with international labor standards and oversees an accreditation program for companies that demonstrate compliance with standards for fair labor and responsible sourcing and/or production. FLA also investigates and works to resolve labor issues in response to formal complaints about companies around the world. This program fills the gap that exists in the absence of a unified set of regulations across formal governments worldwide.

There are hundreds (if not thousands) of similar efforts to structure economic activity so that it complies with indicators we discussed in previous chapters. The net effect is to create a governance process that functions in parallel to the regulation, service provision, tax incentives, and market structures developed by governments. Individuals can help promote sustainability by supporting these efforts, but, of course, not all of them are well conceived. Some are poorly administered, and some amount to little more than greenwashing (see chapter 2). Evaluating an effort requires moving beyond a label and examining which sustainability indicators are being used to certify a product. As such, it is good to have a broader understanding of sustainability when evaluating these nongovernmental efforts. It is also good to keep these efforts in mind as one evaluates the actions of governments themselves.

How else do nongovernmental organizations (NGOs) contribute to governance for sustainability?

The list of examples here is long, so we focus on just a couple related to water quality. Waterkeepers are nonprofit organizations that rely on staff and volunteers to protect and advocate for the bodies of water they represent. Riverkeeper, an organization created to clean up and protect the Hudson River in New York, was the first waterkeeper, but Riverkeepers now take responsibility for thousands of miles of rivers in the United States and around the world. Affiliated with the Waterkeepers Alliance, Riverkeepers are joined by Baykeepers, Lakekeepers, Gulfkeepers, and many others. While the work of each organization differs depending upon location and circumstances, Riverkeepers in the United States represent a good example of how nongovernmental partners work to fill gaps left by underfunded and understaffed state and federal water programs. Riverkeepers monitor water quality through expansive water sampling and testing programs, patrol rivers watching for evidence of water quality problems, and organize volunteers for river cleanups. They also serve as advocates for their rivers, using the legal system where necessary to push for enforcement of federal and state environmental laws.

Another example of collaboration among multiple entities to meet public environmental goals is the South Platte Coalition for Urban River Evaluation (SPCURE). One of hundreds of such organizations around the United States, SPCURE is a partnership of private companies and municipalities in the Denver, Colorado, metropolitan area that hold permits to discharge wastewater into the Platte River. SPCURE members collectively manage wastewater discharges to meet discharge limits and protect water quality. The group also conducts water quality monitoring and modeling to ensure that state water quality goals are met. When considering how such forms of direct action combine with the role that nongovernmental agencies have in standard setting, one sees that collaborative governance contributes to the work of rowing, complementing that done by formal government.

What is the role of government in governance for sustainability?

This question brings us back to government, that is, actions taken by federal, state or provincial, and local entities operating on behalf of the state and subject to democratic oversight. But now we can discuss how governments operate within this space. Governments do have means to compel action that nongovernment authorities lack; however, the use of this power feeds back to the political processes that put elected officials in positions of authority. Governments can often be most effective when they find ways to enable private actors, rather than acting directly themselves. Government agencies may be called in when the certification of private standards becomes overly complex, or where (as with fire codes) there is a need for assurance that standards are, in fact, being met.

Government is also called in when there is a need for rapid or coordinated response, or when threats to sustainability are so large that nongovernmental actors cannot cope: responding to an incursion by foreign invaders, a hurricane, a tsunami, and so on. As we complete work on this book, governments around the world are struggling with the COVID-19 pandemic. The two of us work together while sitting miles apart in our own homes. Our place of employment is locked down, as are many businesses in our community. We cannot visit local restaurants (except to order takeout), and many local stores are closed. Other businesses, like grocery stores, are operating under reduced hours. This same situation is playing out across the United States and other countries. Not surprisingly, in situations such as this, individuals see systems at a different level than governments do. As a result, individual decisions may not reflect what is best for the population as a whole. Questions about whether governments respond to pandemics like COVID-19 in the best way reflect this tension.

Whether they are being asked to respond to business or environmental or social justice concerns (or a whole host of other concerns), governments aren't always effective. They aren't always successful. They aren't necessarily free of corruption. They can't always do the things citizens and residents think they should—or they don't do them well—either because of a lack of knowledge or a lack of money. As noted above, the art of gaining popular support and then making policy constantly reveals unexpected systemic dimensions, even as some ideological rifts appear to be quite sustainable. When governments are involved in hotly contested distributional and social justice issues (discussed in chapter 6), entrenched ideas often prevent them from taking actions that would otherwise appear quite reasonable. (We beg readers' indulgence in not offering an example; there is no way to do it without making half our readership very angry). As political leaders and government employees attempt to make policy, they get so bogged down in the political processes involved in negotiating what to do and how to do it that they lose sight of the end goal. And sometimes governments fail.

Why do governments fail?

What kinds of pressures push a government toward failure? Stated another way, where does one see feedback relationships in these systems? Threats to national cohesion and security may come from illegitimate military activities or high levels of crime. Alternatively, internal strife may arise when a government is no longer believed to be representing all citizens or when sharp divisions among groups within a society lead to oppression, injustice, and violence. The Cuban revolution of the 1950s exemplifies the failure of a government, the authoritarian government of President Fulgencio Batista, who gained power during a military coup, because it was deemed illegitimate, corrupt, and oppressive. High unemployment, poor public infrastructure, divisive economic policies, and evidence of alliances with organized crime (all stocks of one sort or another) were but some of the things that led revolutionaries to ultimately remove Batista from power (feedback to an outflow of legitimacy). In its early years, the new government worked to insure greater social equality, literacy, and public health and to reduce unemployment. However, steps taken by the new communist government to cement power and enforce laws and dissatisfaction with economic and political conditions have led many Cubans to emigrate to the United States and other countries.

One can see similar feedback relationships between stocks related to economic health and perceived legitimacy of a government. A declining economy, indicated by high levels of government debt, high rates of inflation, high unemployment rates, poor business climate, and lengthy recession or even depression, weakens a government's ability to provide services and threatens its legitimacy among citizens. Yet economic prosperity may hide economic inequalities that threaten a government's legitimacy. Venezuela is a country where economic pressures represent a significant threat to government stability. Venezuela's economy, which is almost completely reliant upon oil exports, is particularly vulnerable to economic shocks and market fluctuations in other countries. Recent drops in global oil prices have crippled the Venezuelan economy and led to soaring inflation, unemployment, and poverty. Policies instituted by the current government, combined with corruption and failure to stem rising crime rates, are exacerbating the financial and public health crises within the country. Public protests, many of them not organized by any opposition organization or political party, are widespread, and economic and political analysts have agreed that a transition to a new government is likely within only a few years. Venezuela's contested 2018 presidential election may well be the beginning.

Population growth and associated pressures on natural resource stocks like food and water can seed discontent, as can involuntary exposure of members of society to unacceptable environmental hazards (the resource depletion and pollution problems we discussed in chapter 4). Environmental conditions such as lengthy droughts and associated periods of famine and/or water shortages have been implicated in revolutions and government transitions in France (late 1789–1799), Mexico (1910–1920), Ethiopia (1974–1991), and the Arab

Spring protests of 2010–2012. Climate change will exacerbate these kinds of hazards, but measures taken to address climate change can create hardships for segments of a nation's population as well. Ultimately, when citizens lose confidence in the ability of a government to meet their needs, whether because of corruption, lack of transparency, lack of representativeness, or inability to provide basic protections and services, the government loses its legitimacy. In all these cases, underlying environmental disturbances affect stock and flow relationships in social systems. The final outcome is to weaken the sustainability of the government.

Why are so many governments struggling?

Governments of all types face increasingly difficult challenges. National, state, and local governments operate within economic systems that fluctuate wildly, and economic systems around the world are more and more interdependent. Meanwhile, despite decades-long economic development initiatives, inequities in the distribution of wealth across and within nations are growing. Decision makers operate in an environment in which strengths and weaknesses of governments in far removed places and decisions made by them have unavoidable implications. Governments must make decisions about new technologies despite considerable uncertainty about their long-term benefits and risks. The effects of acts of terrorism, natural disasters, and public health crises are not bound by national governmental boundaries. The economic impacts of investment or disinvestment decisions made by state and local governments spread beyond state and local borders. The level of public distrust in democratic systems is unprecedented, and competition between polarized factions within society to control public discourse and resources is escalating.

From claims of social justice to challenges of pollution, resource depletion, and environmental change, governments at every level face significant challenges in planning for the future while attempting to remedy insults, damages, and injustices of the past. At the same time, at all levels, governments must provide an array of public services and protections to populations that are deeply divided in what they believe is needed and what they are willing to help support financially. In his book *The Honest Broker*, Roger Pielke Jr. describes the problems faced by government decision makers as being one of two general types: tornado problems or abortion problems. His point is that some problems are fairly easily understood, and the solutions are reasonably straightforward and unquestioned. However, other problems defy solution, in part because there is agreement neither on the nature of the actual problem nor on the desired outcome. In the case of a tornado barreling down on a community center, all of those inside can, with a brief bit of research, verify the coming tornado and reach agreement that they should all move to the basement tornado shelter. In contrast, agreement on how to handle the issue of abortion rights is unlikely. Fundamental disagreements on points like whether life begins at conception, the rights of women to make decisions about their own bodies, and whether public funds should support medical procedures that large segments of the public find offensive preclude agreement on whether or not abortion is problematic and what, if any, public intervention is necessary. The difference between these two kinds of problems can be characterized according to the degree of conflict among value systems held by those engaged in the debate and the degree of uncertainty about the nature of the problem and potential responses. When value conflicts are rancorous and uncertainties are high, governing is difficult. Around the world, value conflicts are escalating and scientific and technologic uncertainties are ubiquitous. It is no wonder that concerns about the fragility of some governments are growing. When values conflict and uncertainties are high, questions about steering and monitoring are more likely. Whose interests are favored? Are the costs and benefits of public decisions distributed equitably? How is knowledge to support decisions generated?

Are there indicators of whether governments are struggling?

Yes. This is yet another example of how indicators are used. Since 2005, *Foreign Policy* magazine annually publishes a report describing the risks faced by national governments around the world. This report, produced by the nonprofit Fund for Peace, uses the Fragile States Index to assess the relative fragility of governments based on a series of indicators related to national security, economics, public health and social well-being, and political processes.[2] The premise is that more fragile governments are more vulnerable to failure. Despite general evidence that most governments have become less fragile, the 2018 Fragile States Index report highlights a number of concerns: civil strife in Syria and Yemen, economic pressures in a number of South American and African countries, and political divisiveness in Spain, the United States, and the United Kingdom.

Like the SGI, the Fragile States Index is not used to evaluate sub-national governments, but examples of state governments and city governments that have struggled over the years are easy to find. Many, though not all, of these examples have manifested as bankruptcies, where governments have been unable to carry out their core functions because of inadequate financial resources. However, looking under the rug of bankruptcy reveals weaknesses resulting from corruption, lack of transparency, short-sightedness, and social injustice. In our home state of Michigan, the 2013 bankruptcy of the city of Detroit is a case study in how decades of inequality and institutional racism, poor financial decisions, and political cronyism, buttressed by waves of national and state-level economic and policy pressures, resulted in a city government no longer able to meets its obligations. Further, weaknesses within federal, state, and city governments are behind the failure of the city of Flint, Michigan, to provide a most basic service—a safe drinking water supply—to its citizens. Opaque federal and state regulations, misplaced government priorities, and efforts to deflect blame and disavow responsibility combined with financial exigencies to create the Flint water crisis, and the resulting distrust in government is deep. Yet a renewed engagement of the city's residents—collaboration—with both the steering and rowing activities of Flint's system of governance has grown out of this crisis.

Does collaborative governance help?

We argued earlier that citizens and organizations outside the formal boundaries of government can undertake many of the activities (i.e., rowing the boat) that would otherwise fall to government. There are scores of examples where entities other than formal governments are helping to reduce the burdens (both financial and political) of government agencies charged with providing public services, protections, and opportunities. Federal, state, and local governments have engaged with industry and citizen groups to establish policy priorities, set policy goals, and plan and implement programs of all types. Governments are charged with establishing and enforcing standards to protect public health and welfare, but no government has the resources to fully accomplish such a task. Governments are charged with monitoring public health and welfare, including environmental health, but no government can implement monitoring programs that are expansive enough to respond to the extent of public concerns. Governments are asked to provide a range of public services, but budget restrictions limit their ability to meet this demand. When nongovernment actors share these responsibilities, desired outcomes are more likely. As we noted earlier, they enhance legitimacy, augment resources, and help build knowledge required for difficult, conflict-laden decisions.

Here are two examples—one for how collaboration augments public funds and one for how collaboration reduces conflict. By public funds, we mean the financial resources that governments expend when fulfilling their various roles and providing the public services that have been identified as important. The largest source of public funds is taxation of all sorts. Income taxes, property taxes, and sales taxes are the big three. Former US Supreme Court justice Oliver Wendell Holmes Jr. famously said that taxes are the price we pay for a civilized society. Taxation is also one of the most vilified governmental activities. This is not the place for a lengthy discussion of appropriate levels of taxation or questions about whether tax structures are fair. The important point here is that the money to enable governments to meet their obligations has to come from somewhere. Examples abound across the globe of governments that, for various reasons, do not have enough money to provide the most basic public services. Even more examples can be found of governments, especially in wealthier countries, states, or cities, that are striving to meet a broader set of social goals with limited funds. In both cases, collaborative efforts have evolved to find ways to channel additional financial resources into supporting publicly desired activities that governments have been unable to support.

New York City's park system is a good example of how local government partnered with the private sector and concerned individuals to supplement public resources with private funds to meet its mandate. As a result of the financial crisis weathered by New York City during the 1970s, the city parks department faced an impossible task of managing all its parks and associated facilities. The financial exigencies left Central Park, an emblem of New York City's gilded past, with dilapidated grounds, crumbling infrastructure, and abandoned structures. City residents and visitors alike feared for their safety within the park. Over the decade of the 1980s, a host of partnerships were developed to harness private resources to rehabilitate the city's parks. One such partnership led to the creation of the Central Park Conservancy. The conservancy sought gifts from wealthy New York residents and used those funds to refurbish Central Park, working with the city and an army of volunteers. Now, the conservancy and the city manage the park together, but the conservancy raises the funds for the park's nearly

$80 million annual operating budget, and almost all park staff are conservancy employees.

In recent years, the concept of civic crowdfunding has garnered considerable interest and is seen by many as an opportunity for citizens to contribute financially to public projects and programs they support. Most examples of civic crowdfunding are cases where citizen groups or civic organizations have initiated local, small-scale projects like building community centers, revitalizing parks, fighting blight through beautification programs, and the like. The jury is still out on how successful these kinds of collaborations will be in the long run; this type of broad-based civic action requires careful organization and committed leadership. Civic crowdfunding has its critics, though, and they make arguments much like those made by early critics of New York City's parks funding model. A principal concern is whether these kinds of funding models unfairly favor the interests of wealthier individuals who can better afford to participate when it comes to setting public goals and establishing public priorities.

We have discussed how social conflicts can threaten the stability and functionality of governance processes. They can also topple governments. From there, we have discussed how questions about the environmental sustainability of manufacturing, energy use, and processes contributing to climate change can provoke social conflict. There is, in short, a potential connection (a feedback) between challenges to environmental sustainability and challenges to the sustainability of good governance processes. As the ecological systems and natural-resource base that support our way of life become threatened, these threats trigger an increase in the disaffection, strife, injustice, and anger that weaken a government's ability to govern. Thus a key question for sustainability is whether nongovernmental governance processes such as standards, NGO activism, and shifting the funding for sustainability away from the tax base can reduce the social conflict (we might think of this as a stock) that creates destabilizing political feedback. We believe that the answer to this question is yes.

One example of how partners outside of formal government work to reduce conflicts that might otherwise require government or legal intervention is found in Canada's Ombudsman for Banking Services and Investments. This program was created in 1996 to resolve disputes between consumers and members of the financial services industry. It is not part of the government. It is funded by the industry but is a separate, independent entity. Having this third-party organization available to consumers and to banks and other financial firms reduces the burden on government regulators and the number of disputes finding their way into the legal system.

Another example can be found in East Lansing, Michigan, home to Michigan State University (MSU). Following a series of alcohol-fueled clashes between university students and city officials in the late 1990s, relations between the university and city were strained. Many citizens grew to resent the student presence in the community, and local support for the university was waning. The Community Relations Coalition (CRC) was created in late 1999 as a partnership between the city of East Lansing and MSU; the overarching goal was to create a more integrated community. The CRC, with the motto "We All Live Here," is a nonprofit organization with a board of directors that includes students living on campus and off campus, representatives of city government and police, MSU administration and police, and city residents. MSU student interns serve as ambassadors to students and permanent residents and work closely with neighborhood partners to address community issues, foster collaboration, build relationships, and share information of broad interest.

How does climate change threaten the sustainability of governance?

The SGI (discussed above) includes indicators that rate governments according to their level of participation in international efforts to reduce climate-forcing emissions. This reflects the view that failing to participate in activities that increase global environmental sustainability makes governments themselves less sustainable. Pressing more deeply, we offer some speculations but concede that this is a question needing both research and thoughtful consideration. Governments in Europe, especially Germany and Scandinavian countries, rank high on this component of the SGI, while the United States is notoriously unwilling to sustain a political commitment to remediation. (Again, see *Climate Change: What Everyone Needs to Know®* for more details.) One could regard domestic support for international cooperation as a stock in each country and examine how this stock creates feedback to a political stock that is easy to measure: votes. We surmise that the climate support stock is already higher in Germany than the United States, and that when governments are successful in linking economic development activities (such as public funding for the power grid) to efforts that reduce emissions, this feeds the inflows for both votes and support for climate policies. The German parliamentary system may also play a role, as minority parties (such as the Green Party) can achieve goals by partnering with other groups.

In the United States, a sizeable segment of the electorate is skeptical of climate projections, a stock that may also be connected to skepticism about the role of government in general. The actions that the SGI watch to measure support for international cooperation may trigger feedback that increases climate skepticism and mistrust of government—exactly the opposite of what may be happening in Germany and Scandinavia. The winner-takes-all feature of the American voting system means that this skepticism can sometimes put the party that rejects international cooperation over the tipping point in the votes stock, putting them in power. But why would this make the US government less sustainable in the sense of being less able to continue? The answer to this question (again speculative) might include feedbacks to still more stocks, such as US influence in other areas of trade and international agreements.

Relevant stocks might even include the goodwill extended to American tourists or businesspeople abroad. And, of course, when the effects of climate change become visible, one would not expect politicians or regimes who have opposed mitigation efforts to fare well politically.

These stocks, flows, and feedbacks are difficult to measure objectively. The challenge of developing a quantitative model for sustainability of governance certainly contributes to the difficulty of evaluating governance in terms of sustainability. At the same time, the potential for a large-scale failure, as well as the injustice of imposing the costs of US emissions on those least able to bear them, may do more to expose weaknesses in the sustainability of a governance system than the myriad domestic failures that have been tolerated for decades.

How do pandemics threaten the sustainability of governance?

Plague is one of the four horsemen of the apocalypse. As with climate change, public health crises expose weaknesses in the sustainability of key governance processes. Governments' reactions to the COVID-19 pandemic in 2020 reflect, in part, their planning and preparations for such an eventuality. That, in turn, is affected by past experiences with pandemic disease and by the influence of public health professionals on governance. Leaders of several countries were criticized for either a lack of long-term planning for how to respond to a pandemic or a failure to take decisive action quickly enough. Both criticisms call into question issues of sustainable governance if the SGI attention to governance processes and executive capacity is accurate. A key piece of the executive capacity indicator looks at the degree to which leaders heed the advice of scholars and consult with social and economic interests. In contrast to outbreaks of the SARS virus and Ebola, the COVID-19 pandemic exposed feedback between public health policies and economic activity that are not commonly accounted for in government planning.

As we remarked earlier, governments' decisions commonly determine who will bear what costs. The same is true here. On the one hand, many governments acted to limit the spread of the virus—hence the business closures and orders to "stay home and stay safe." On the other hand, governments acted to offset some of the financial hardships that businesses and workers experienced and the larger economy-wide threats and weighed the relative financial risks of extending business closures with the health risks of not doing so. Calls by some financial interests to "reopen the economy" met with skepticism from segments of the population concerned about who would be affected positively or negatively by such a move. If the economic engine is fueled by workers who are more vulnerable to the virus or who have inadequate access to healthcare, the unpopularity of ending isolation and resuming normal economic activities would not be surprising. In regions where access to healthcare is problematic as a matter of course, overloading the healthcare system is one more point for erring on the side of caution by keeping people distanced from one another. We noted earlier that the sustainability of a governance system is challenged if it is unable to provide basic services, protections, and security. Pandemic disease represents a key threat to the sustainability of governance practices that do not account for linkages across subsystems or that fail to plan for low-probability, high consequence events.

What is the take-home message?

As we said at the outset of this chapter, every book about sustainability discusses government, but governance systems are seldom part of larger conceptualizations of sustainability. Arguments about what governments should do to support businesses, protect ecosystems, ensure environmental quality, advance economic development, and pursue social justice are common, but attention to the ability of governance systems to do these things is less common. Oddly, people who have gained sophistication in using systems thinking to understand sustainability in business, ecology, environmental quality, and the social realm seem to forget everything they know when it comes to government and governance. The important message from this chapter is that attending to some key elements of governance systems—stocks, flows, feedbacks—as well as system hierarchies and resilience helps us understand what contributes to their sustainability.

The examples described throughout this chapter suggest critically important stocks that may destabilize a governance system if they are too large or too small. Too much oppression, economic inequality, unemployment, corruption, or violence introduces fragilities into governance systems. Too few basic services, shortages of food and water, and too little government transparency or public financial capacity also put governance systems at risk. Feedback relationships may amplify these problems. A lack of economic opportunity contributes to economic inequality. A lack of transparency breeds concerns about government corruption. The effectiveness of national, state, and local systems of governance are closely tied to one another, and governmental decisions made in faraway places can reverberate through governance systems close to home. Participation of nongovernmental organizations, businesses, and individuals in governance processes enhances the resilience of governance systems. And when governments fail to listen and respond to public concerns, this communication breakdown threatens that resilience.

This is a good time to reiterate what we said in chapter 1 in answer to the question "Is sustainability always about the environment?" The answer was no, but when people turn toward thinking about government, they tend to emphasize policies intended to curb the impacts of pollution and resource depletion. As climate policies have entered the discussion, systemic processes have gotten more emphasis, but governance processes are still treated as if policy decisions do not initiate feedbacks that reverberate throughout the total system. At the end of the day, sustainability writ large is about large-scale social goals, and our systems of governance, including formal government, are the principal way in which societies pursue those kinds of goals. It is crucial to understand the process of governance in terms of the stocks, flows, and feedbacks that allow it to continue—that is, its sustainability.

8

SUSTAINABILITY IN SCIENCE, EDUCATION, RELIGION, AND THE ARTS

Why discuss sustainability in science, education, religion, and the arts?

The things that people do in science and education support economic activity, environmental quality, and governance, while religion and the arts lend meaning and purpose to a community's way of life. Each of these four areas has some form of enlightenment, edification, or growth in knowledge and awareness as a mission or organizing purpose. As such, it is appropriate to accentuate the ways in which teachers, researchers, spiritual leaders, and artisans help us appreciate the essentials of sustainability.

Sustainability becomes relevant to these four domains in two ways. On the one hand, the systemic interconnection of the economic and environmental spheres exposes vulnerabilities in the way all of us do business as a society. One can ask how scientific research, schools, churches, and creative practices increase our collective capacity to reduce those vulnerabilities. That question would probe how each of these domains can help us become more sustainable. On the other hand, someone could ask, what does it take for practices in science, education, religion, and the arts to continue? For example, the things people do in all four of these areas require some amount of monetary support. One could try to understand the systems that provide that support. They might include private contracts and grants in the case of science, taxes or tuition payments to schools, donations to religious organizations, and a mix of for-profit and nonprofit support for the activity of writers, musicians, filmmakers, painters, sculptors, and other creative individuals.

In other words, a person can ask what people in each of these spheres are doing to promote sustainability, or they can ask what it takes to make these four kinds of practice sustainable in themselves. We do a bit of both in this chapter, but unlike the chapter on governance, where we emphasized the sustainability of governance processes, here we stress things that scientists, educators, religious practitioners, and artists do to help everyone improve the sustainability of our social institutions and our general way of life.

How do science, education, religion, and the arts address sustainability?

The goal of making all aspects of contemporary society more sustainable influences thinking and behavior throughout many different domains of life. The four domains we discuss in this chapter share the goal of increasing knowledge, awareness, or capacity, but they obviously differ dramatically in both their approach to this goal and the kind of awareness and capacity they aim to enhance. Not only did sciences such as ecology and systems engineering give rise to key ideas in sustainability, but the organization and conduct of scientific research has undergone important changes in the era of sustainability. These changes in science, technology, and engineering have in turn led to revisions in educational programs at the college level. In religion, reflection on the religious basis for conserving and protecting nature influences the way that faith traditions are practiced and, in some instances, the very meaning of spiritual enlightenment. The arts have responded with creative ways to bring issues of sustainability to broader public attention and to inspire individuals toward collective action or personal choices that address environmental and social challenges.

We pose a series of questions related to each of these four ways in which people pursue learning and increase their capacity for action. Each question could be examined at a finer level of detail. One could, for example, go into the way that physicists study sustainability and distinguish that from the work of economists. One could examine how sustainability has resonated within different religious faith traditions, and there are so many different forms of creative activity in the arts that it boggles the mind. There are also professions and vocations that are not obviously covered under these four headings: healthcare, tourism, recreational activities, and sporting events, for example. Sooner or later we would be too deep into the weeds. Here we emphasize questions that we believe will help readers understand and appreciate efforts in these four domains.

What is sustainability science?

Sustainability science is scientific research designed to identify the unsustainable practices that threaten ecosystems and to develop alternatives that preserve the integrity of ecosystem services. Both natural and social sciences play key roles in disclosing the importance of sustainability. Up to now, research conducted under this banner has stressed environmental dimensions of sustainability. There has certainly been research on the financial sustainability of organizations or the stability and resilience of governance institutions, to point out two topics that we discuss in this book. Nevertheless, the term *sustainability science* is largely reserved for studies that emphasize human impact on environmental systems. For example, sustainability science on business activity has stressed life-cycle analyses and resource use, rather than the relationship between sustainability and profitability that we discussed in chapter 2.

Sustainability science didn't appear out of thin air. In chapter 3 we saw how the concept of sustainable yield arose in connection with basic ideas in forest management and population ecology. In more recent times, environmental scientists have generated the concept of planetary boundaries (also discussed in chapter 3), while models of the interaction between greenhouse gases and the atmosphere illustrate cycles that give rise to dramatic change in climate systems (discussed in chapter 4). Economists studied how markets would adapt to the increasing scarcity of finite resources and identified how potentially renewable resources could become threatened. In short, science is fundamentally responsible for our ability to understand both social and environmental sources of vulnerability from unsustainable practices.

In 2004, the US National Academy of Science created a special section in their signature journal *Proceedings of the National Academy of Science (PNAS)* for sustainability science. The goal of this prestigious organization was to both recognize and encourage new scientific research that would be directly applicable to sustainability and that would aid all manner of decision-making. This problem-solving orientation was a break from the way that scientific disciplines traditionally organize and prioritize research. Sustainability science thus does represent a departure from the way that classic disciplines (physics, chemistry, biology, economics, psychology, for example) are organized.

How is sustainability science different from applied science?

Science has long recognized a distinction between basic and applied research. In all the scientific disciplines, curiosity about the workings of phenomena provides an important impetus to inquiry. The aim is to produce more complete and accurate explanations for the events and processes that a given discipline studies. Such curiosity-driven research is classified as basic science. Basic research in physics and chemistry has given rise to new and more powerful models of the material world. Basic research in psychology leads to more sophisticated understanding of memory, perception, and human behavior. Advances in these fields have been adapted to specific problems or technologies and led to revolutions in transportation, agriculture, medicine, and the like. The research needed to adapt basic science for practical application is called applied science.

Many types of applied science are considerably more lucrative than basic science. With important exceptions in agriculture, public health, and natural resource management, private investors pay for and manage applied research. Patents, innovations, and new technologies generate financial returns to the scientist (or the scientist's employer). Basic science may support these inventions and products, but the findings of basic science are freely available to everyone. Basic research produces publications in scientific journals rather than a marketable product. Throughout the twentieth century, universities and government or nonprofit institutes organized their science programs to favor basic research. The assumption was that the financial rewards for applied research provided a sufficient incentive for others to do it. Universities accepted a responsibility to serve the broader public good, rather than seek profits, and favoring basic science was a part of this mission. After World War II, the US government also followed this model, establishing research agencies such as the National Science Foundation (NSF) and the National Institutes of Health (NIH). Here, too, basic research is favored. An NIH scientist might work on the basic biology or mechanisms of a disease process, but development of drugs or devices to cure it is left to the private sector.

The problems of unsustainable practice do not lend themselves easily to a model of profit-driven research. The problems are complex and require coordinated action on many fronts, rather than development of a simple product that would resolve everything. As with applied science, underlying theories need to be adapted and specified in order to address a problematic situation, but there is little reason for the private sector to expect financial rewards for conducting much of this research. The solutions may not involve anything that can be sold to recover the costs of doing the research. At the same time, the recognition of a problem determines what type of research needs to be done; the research is problem-oriented rather than curiosity-driven. Hence, research for sustainability does not fit well into the organization of basic science. The suggestion that research on solutions to social or environmental problems should override research that addresses unknown areas at the frontiers of each discipline is foreign to much of the practice of traditional disciplines. Such studies are difficult to publish in prestigious journals, and many scientists question whether they will be rewarded for doing such research. The idea of sustainability science was born from the recognition of this weakness in the organization of research in the natural science disciplines.

Sustainability science followed several different pathways in its development. Complexity science emphasizes new techniques in mathematical modeling to better describe the complex interdependencies of the social and ecological systems on which life depends. These studies often require contributions from many different scientific disciplines, so interdisciplinary science is a complement to other types of sustainability science. Finally, problems of sustainability are often ill-structured, with different interest groups preferring mutually incompatible responses. They are known as *wicked problems*, and sustainability science has developed new methods to cope with them. In one approach, sustainability scientists have developed tools for studying coupled natural and human systems.

Emphasis on participatory research is a second approach. Classic forms of curiosity-driven science follow the interests of the scientific community and rely solely on the knowledge base that exists within a scientific discipline. However, nonscientists may be much more aware of the problems that need researching than scientists, and they may have knowledge that scientists can build upon in developing a response. What is more, implementation of scientific results often requires action by non-scientists. New methods for involving citizens in the design of a research project, the collection of data, and the analysis of results make scientific research more responsive to the problems that need resolution. They also involve people from all occupations, ethnicities, and social groups more thoroughly. This helps to broaden the knowledge base and reduces the need for a separate effort to translate the results of a scientific study into concepts that non-scientists can use.

Complexity science and participatory research have emerged as distinct specializations in their own right. What is more, both methods of sustainability science require integration of knowledge from classical scientific disciplines—curiosity-driven science. Each of these research approaches involves its own ways of collecting and analyzing data, and the results they produce can be difficult to assemble into a unified approach to problem solving. Thus a further task in sustainability science is to develop ways to improve the integrative capacity of science. Sustainability science thus requires new methods of inquiry that cross disciplinary boundaries.

What are wicked problems?

Horst Rittel and Melvin Webber formulated the idea of wicked problems for urban planners in the 1960s. They were pointing out the difference between "tame" problems that could be solved with a technical analysis and those that exceeded the technical skills of traditionally trained urban planners. Rittel and Webber identified ten characteristics of wicked problems:

1. Wicked problems have no definitive problem definition.
2. There is no way to tell when research on a wicked problem is complete.
3. Solutions to wicked problems are not true or false but good or bad.

4. One cannot be sure that one is making progress on a wicked problem, or that it has been solved.
5. Wicked problems defy attempts to learn by trial and error.
6. The number of possible responses to a wicked problem is effectively infinite.
7. Every wicked problem is essentially unique.
8. Every wicked problem is embedded within other wicked problems.
9. Whether the solution of a wicked problem is good or bad depends on the perspective one takes.
10. The planner has no right to be wrong.

This list (which we have modified slightly) has generated a lot of discussion among planners and sustainability researchers. We emphasize just a few points that are particularly pertinent to thinking about sustainability.

For many researchers in sustainability, the first characteristic of wicked problems—lack of a definitive problem definition—is the most important one. When there is general agreement on what the problem is, teams of researchers can be assembled to identify the possible solutions to it. When people disagree on what the problem is, it is difficult, if not impossible, to be sure that one is making progress, because the criteria for progress will be in the eye of the beholders. Some sustainability researchers prefer to call these ill-structured problems, because the word *wicked* seems to imply a moral judgment where none is intended. Others have said that even using the word *problem* implies too much coherence and agreement: people should instead refer to wicked problems as imbroglios. They are more like a bar fight than a reasoned argument.

There is a difference between complex problems and wicked problems. Systems with many components behaving in intricate and difficult to predict ways present extremely challenging problems for scientific researchers. Many natural phenomena exhibit such a high degree of complexity that they continue to defy the creation of accurate models that would predict their behavior. Although engineers have developed powerful tools for analyzing the behavior of liquids and gases, these models are not capable of accurately predicting when flows will become turbulent. Although this problem has a very high degree of complexity, it is not wicked. Everyone agrees that a mathematical model capable of predicting and controlling turbulent flows of air or fluid would count as a solution to this problem.

Finally, in saying that the planner has no right to be wrong, Rittel and Webber were stressing the irreversibility and high stakes that are involved in most wicked problems. Attempts to ameliorate or manage a wicked problem usually involve risk-taking and impose costs on some parties. Getting involved in a wicked problem will also have an impact on the situation, changing the initial conditions or structure of the problem and preventing a do-over. As such, any attempt to intervene in wicked problems is fraught with ethical responsibilities. Researchers and policymakers must consider which parties are most at risk from attempts to improve the situation and determine the fairest distribution of costs and benefits.

In some ways, one can compare the characteristics of wicked problems delineated by Rittel and Webber to Roger Pielke Jr.'s distinction between the two kinds of governance problems (described in chapter 7, if you skipped that part). When there is general agreement on the nature of a problem and the desired outcome of an intervention, the politics of governance is uncomplicated. These would be, for Ritter and Webber, tame problems. But Pielke notes that values conflicts and technical uncertainties are high for some of the most enduring political questions (his example is abortion). Agreement on defining the problem is lacking, and views on what would count as a solution differ. These are wicked problems. Agreement on what the outcome of some intervention ought to be and whether a desired outcome has been achieved requires some common set of values among those affected. Wicked problems bring with them considerable uncertainties, whether in terms of the causes of problems, the effectiveness of efforts to resolve them, or the irreducible ways in which actions reverberate through natural or human systems and introduce myriad new problems.

What is interdisciplinary science?

Interdisciplinary science is a catch-all term for several different types of novel research approaches for sustainability. Multidisciplinary research requires coordinated input from different disciplines. As an example, James Watson, Rosalind Franklin, and Francis Crick's discovery of the structure of DNA required inputs from biology, microscopy, and physics. In addition to contributions from several disciplines, interdisciplinary research requires close interaction among researchers from different disciplines because the definition of a research problem cuts across disciplinary boundaries. Agricultural science, for example, often requires additional steps to make the separate studies of crop scientists and farm management economists relevant to a farmer. There are also transdisciplinary problems that lie in between areas of knowledge where theory is well developed or that require both scientific methods and more practical or humanistic arts in order to be understood or addressed. The term *interdisciplinary inquiry* or *interdisciplinary science* is commonly used for all these situations.

Scientists debate what it means to do this kind of work: many find it difficult to adapt to new ways of thinking. Most researchers are trained within the confines of specific disciplines, each of which comes with its own vocabulary, body of theory, and accepted research methods. Universities and research institutes are organized into administrative units by discipline. Career advancement is tied to disciplinary contributions. Even when scientists set out to tackle wicked problems through interdisciplinary work, they face barriers. Learning the language of different disciplines is a barrier. Different disciplines have different standards and methods for data collection and analysis, and they publish in different outlets. In recognition of these challenges, we now have research on how to do research.

It studies methods for forming and running teams of scientists from different disciplines. More and more, universities are looking for ways to foster cross-disciplinary work that can address the wicked problems of sustainability.

What is complexity science?

A system is complex when it is intricate or complicated. Complexity science responds to the complexity of unsustainable practices by developing theoretical constructs for guiding research on them and creating new tools in mathematics for modeling highly intricate and complicated processes and practices. Research on complexity begins by cataloging all the different ways that complexity is measured or analyzed in distinct areas of scientific inquiry. For example, in information technology, the complexity of a sequence is measured by the minimum number of bits in a program designed to produce it. Complex adaptive systems are defined by the number of units interacting within the system, the presence of feedback, and the nonlinearity of interactions between the system and its environment. More generally, complexity is considered organized if the interaction of parts can be tracked by a mathematical model; it is disorganized when the interactions are effectively random. Charting the ways in which complexity is defined or addressed in different domains is crucial for interdisciplinary studies of intricate and complicated phenomena.

Complexity has emerged as a phenomenon of interest even within basic sciences. Among the basic sciences, physics and economics have developed increasingly sophisticated mathematical approaches to explain the behavior of phenomena that each discipline studies. Chaos theory, for example, is a field of mathematics that studies the behavior of dynamic systems that are highly sensitive to their initial conditions. Chaos theory improves the accuracy of weather predictions by assigning probabilities to scenarios that can emerge as extremely complicated weather systems interact. Climate modelers apply chaos theory to assign probabilities to much longer-term scenarios of underlying conditions that influence weather on a daily or weekly basis. In sustainability science, these powerful tools for coping with systems that have many parts interacting in highly intricate ways are applied within interdisciplinary contexts and to the integration of multiple knowledge domains.

What are coupled human and natural systems?

By the year 2000, there had already been several centuries of research on the behavior of natural systems such as the atmosphere, oceans, forests, savannas, and wetland ecosystems. Biophysical models for energy flows were developed and tested, and population models for predator-prey relationships were available. No one would have said that these phenomena were exhaustively understood, but quite a bit was known. Similarly, economics, psychology, and the social sciences developed models for understanding and predicting human behavior. These models were applied to monetary policy, business management, and education, with significant degrees of success. However, very little work attempted to pull findings from the sciences of human behavior together with findings from studies of the natural environment. It is obvious, though, that problems of sustainability originate in the way that human beings use or otherwise affect natural systems, including the atmosphere, oceans, and terrestrial ecosystems.

Researchers in the field of coupled human and natural systems build scientific models that integrate what is known about human behavior into biophysical models of stocks and flows of natural systems. In the most basic cases, this involves connecting feedback from human decision-making to the flows of harvest from a forest or a fishery. But it is also possible to model feedback from the natural system to the social system. For example, as stocks of economically important natural resources (such as fish or fossil fuels) decline, prices increase and send feedback through the economic system. Coupling what social scientists already know about human behavior and socioeconomic systems to what biologists know about ecosystems is an obvious—but technically challenging—first step in sustainability science.

Coupled human and natural systems models fill an obvious and important gap in how basic and applied science of the twentieth century addressed problems of sustainability. The systems that matter to sustainability involve biophysical processes and human behavior working interactively. Models that did not reflect this interaction were too simple to have much power in identifying plausible strategies for improving sustainability. While coupled human and natural systems modeling is an important response to the actual complexity of the regional and global systems that impact sustainability, it is less clear that this approach can address wicked problems. The models being coupled were not developed to be responsive to the ambiguity, uncertainty, and high stakes that characterize wicked problems. Coupling economic and ecological models gives scientists a more comprehensive view of the complexity of human and natural system interactions but does not of itself represent a strategy for dealing with the indeterminate nature of sustainability as an ill-defined and contested problem.

What is participatory research?

Defined broadly, participatory research is any form of science in which non-scientists play a central role in defining the problem, developing research methods, and implementing research findings in a practical setting. A hallmark of participatory research is that non-academic partners have a stake in the research problem; the research addresses a clear, real-life problem, and stakeholders want to be part of doing something about it. Participatory methods have been around much longer than sustainability science. Applied researchers in food, medicine, and other fields developed methods for interacting with beneficiaries of their work. However, these traditional forms of applied research seldom include stakeholders as active partners in developing and agreeing on a research approach. Applied research in agriculture, public health, and natural resources benefits large numbers of people, but researchers commonly go to the field with a particular formation of the problem already in mind. Researchers have learned that the kinds of problems invoked by sustainability call for more fundamental involvement of those with stakes in the problems.

It is difficult to overstate how this minor shift has influenced science. A health researcher might draw upon doctors' knowledge of a disease to research drugs or therapies; a plant breeder may talk with farmers to understand problems with drought or pests. In these forms of applied research, the scientist shapes a research process that responds to stakeholder problems. In contrast, the complexity and lack of structure within planning and management disciplines could not be addressed simply by developing a product or procedure and then teaching people how to use it. Applied researchers in planning disciplines and business management have long faced problems mired in controversy and where all potential responses are plagued by uncertainties. In these domains, stakeholders need to be involved in a learning process that enables them to cooperate with one another in effecting solutions. In this type of research, scientific training can aid in defining technical aspects of the problem and suggesting possible responses, but there is no research result—no drug or plant variety—that researchers can hand over to stakeholders. Approaches that enable more active participation from stakeholders have been adapted to wicked problems in sustainability science because methods for working with community partners have a track record in management and planning,

Participatory methods call for an approach that involves non-scientists in the early stages of problem definition. In traditional applied science, participants have common interests and goals; hence they are likely to agree on a basic problem definition. However, in some applied areas, such as management of water or other natural resources, the potential for multiple uses creates a situation with differing perspectives, a feature of wicked problems. Just as urban planners were encountering ill-structured problems in the 1960s, researchers on soil management, water quality, or the use of public lands have long worked to meet the needs of a public with diverse perspectives on how a resource is best used. Key methods involve finding ways to identify and assemble groups of stakeholders and engage them in a process of consultation and deliberation so that conflicting visions or interests are negotiated in a fair and impartial process. Also, bringing stakeholders' multiple ways of knowing (ways of creating and applying knowledge) into the research process is critical, as is recognizing the importance to stakeholders of resolving the problem.

Researcher and stakeholder partnerships have developed strategies for convening groups to develop responses to wicked problems in many different disciplinary and organizational settings. The methods and tactics for participatory research go by many different names. These include soft systems research, design thinking, scenario analysis, the five disciplines, and critical systems theory. Each approach has unique features, but what unites them is the goal of bringing diverse and potentially disagreeing parties together to jointly identify how research tools and scientific or design methods can make improvements in the situation at hand.

Is sustainability a wicked problem for science?

Yes and no. When everyone agrees on the key stocks, flows, and other elements of a system, as well as what the system is supposed to be doing, determining whether or the degree to which system processes can continue to operate is a technical problem. Measuring the rate at which trees can be cut in a managed forest, for example, is a complex but tame problem. The basic flows in this system are determinable factors in the growth rate of trees, though complexity enters due to variability in the weather and the potential for disease outbreaks or infestations of insects. This problem becomes wicked when the boundaries of the system are disputed or when multiple users have different goals and different levels of control over using the system's resources. State and federal forest resource management agencies were confronted with the wickedness of problems when tasked with implementing multiple use management strategies: recreation, watershed, and wildlife values being given equal footing with timber values. In theory, measuring the rate at which fish can be taken from an ocean fishery should be very similar to calculating the rate at which trees can be taken from a managed forest. Fishery sustainability is a wicked problem because people do not necessarily agree on what the problem is. For fishers, it may involve maintaining incomes or a traditional lifestyle, while for seafood wholesalers the system would not be sustainable unless it could deliver a steady stream of product to retailers. Environmentalists likely have still different goals. No one has the authority or ability to control all the parties affecting the fishery's stocks and flows, which, when combined with the fact that fish move around, makes it hard to describe the system boundaries.

When sustainability is expanded into something like the sustainability of our current way of life, the wicked dimensions become more obvious. The different ways of making sense of sustainability that we have discussed throughout the book create scientific research opportunities that overlap and interact in surprising ways. Water specialists find themselves needing to work with economists and communication specialists (i.e., interdisciplinary research). Non-scientists may have important information or may need to be part of the research process in order to test solutions (i.e., participatory research). Researchers need to build models that draw from biophysical and social science (i.e., coupled systems). Scientists may need to revise some of the principles they were taught to do this, and they are uncertain about how to proceed. In all of these respects, sustainability is a wicked problem for researchers, and we have not even begun to talk about complexities that emerge when addressing the way that different levels in the hierarchy of natural, social, economic, and political systems interact. What is more, getting it wrong is not an option. If the environmental systems that support our way of life collapse, the human and social costs will be staggering. In the worst-case scenarios, the planet earth becomes uninhabitable. At this level, there are no do-overs for sustainability. Sustainability researchers have devoted significant time and energy to identifying the best strategy for organizing research in sustainability science.

What is sustainability education?

In a nutshell, sustainability education is the incorporation of all the various elements of sustainability science into educational programs. There are at least two dimensions to sustainability education. First, sustainability science has stimulated changes in school curricula, particularly at colleges and universities. Undergraduate courses and majors are being developed that take sustainability as the main organizing concept. Second, adult educators are also starting to reflect sustainability themes in programs for workforce improvement, retraining, and lifelong learning. Museums, films, and other forms of informal education are also included in these new types of sustainability education.

How do schools address sustainability education?

Starting around 2000, many colleges and universities began to restructure curricula and administrative organization to address sustainability. They created new majors, new departments, and, in some cases, even new schools or colleges or reorganized the entire university. One reason was to facilitate interdisciplinary collaborations. Another was to develop new courses and programs of study less tied to the old models of discipline-based education in fields such as biology, chemistry, economics, or psychology. New courses and majors were created to both integrate all these fields and select those aspects of the fields most relevant to the challenges of sustainability. Sustainability education programs incorporate systems thinking or systems analysis into their training programs. In addition to science training, sustainability programs also include coursework in environmental philosophy and humanities, as well as the creative arts and design.

Some sustainability education programs are little more than relabeled programs in environmental science or environmental studies. What is more, as resilience has moved to the forefront of public discussions of sustainability, the very idea of sustainability appears to be wearing thin to some observers. There is also concern about what jobs await students in sustainability programs after they graduate. A lack of clarity about the content of sustainability education has slowed the creation of employment categories that have sustainability as a focus. Nevertheless, programs in sustainability are extremely popular among undergraduates, generating strong enrollments. Schools are just beginning to adapt sustainability education for elementary and high school students, searching for elements of sustainability curricula tailored to younger learners. Many educators have started to emphasize experiential learning through gardening, recycling, and energy conservation, involving students in the design and implementation of activities intended to promote sustainability. When these programs are well executed, they expose students to complex systems and give them practical experience with the ways that stocks, flows, and feedbacks can affect their ability to meet key goals. These activity-based forms of education in schools overlap with many of the approaches being taken by adult educators who work with people not enrolled in a formal, school-based educational program.

What is adult sustainability education?

Classroom teaching is important, but many people above the age of eighteen (and plenty more above the age of twenty-five) will never set foot in a classroom again. Are there things that they should know about sustainability? Well, we wouldn't be writing a book with the title *Sustainability: What Everyone Needs to Know®* if we didn't think the answer is yes. In one sense, the book you are reading is an example of adult sustainability education: the format of questions and answers, the absence of footnotes, and (we hope) the informal writing style are all intended to help someone who is not reading our book in order to pass a test. At the same time, adult education encompasses a much bigger range of activities and products than reading material.

Museum experiences are among the most effective forms of nonformal adult education. Whether the museum uses a traditional walk-through display or a more activity-based format with interactive projects and experiences, adult visits (often with children) to science, history, and even art museums provide key opportunities to convey information about sustainable systems and the threats to our contemporary way of life. In addition, museums, colleges, and local school systems are experimenting with new venues such as "science shops" or "science cafés" where people can come to learn and talk about the underlying science of sustainability. These methods join with the tried-and-true speaker series: public talks on sustainability at libraries, church gatherings, or anywhere that a group of listeners can be assembled.

The significance of all this lies in the nature of sustainability itself. As a wicked problem involving complex systems, sustainability is not something that will be achieved by experts working behind the scenes. It is not like when you call a plumber and then go see a movie while he or she fixes the leak. Moving toward more sustainable practices requires action on the part of everyone. It needs to be pursued in many venues. In the rest of this chapter, we consider two settings where people can learn more about sustainability and work with others to promote it: religious activities and the arts.

How is sustainability incorporated into religion?

The answer to this question requires a little setup. The medieval historian Lynn White Jr. published an article in 1967 entitled "The Historical Roots of Our Environmental Crisis." He placed substantial blame for consumptive and destructive human use of natural resources on doctrines common to Judaism, Christianity, and Islam. White said that these religions had portrayed nature as a gift from God to humanity, leaving humankind to use and ultimately exploit the gift however they chose. Doctrines of salvation in a life after death seemingly relieved believers in these faith traditions of any reason to concern themselves with the preservation or health of the earth. White's article spawned a debate among religious scholars over the extent to which Western faiths encourage exploitation, rather than stewardship, of nature. It has had enormous influence in schools of theology, if not in local religious communities.

White's article was consistent with other work in philosophy, history, and other humanities fields that stressed the protection of nature as the primary goal of environmentalism. National parks and wilderness areas were seen as places where ecological processes could play out unaffected by interference from human beings. Environmental ethics emphasized the intrinsic value of these natural areas and suggested that any human use of them would contaminate or pollute. Early studies in ecology also emphasized a conception of nature in which human beings were completely uninvolved. Many environmental writers and scholars in the humanities followed this line of thinking, creating an approach that continues to be extremely influential in environmental studies.

The idea of an uninhabited and untrammeled wilderness was especially influential in North America. As White argued, European settlers had viewed the continent as empty, and they saw the lack of agricultural or industrial development as a waste of God's gift to humanity. However, numerous tribes and bands of Native American peoples, representing many distinct language groups and civilizations, populated North America before the arrival of Europeans. Agriculture and hunting or fishing were widely practiced. Planned burns of brush and timber determined how stocks and flows of flora and fauna interacted in forests and grasslands. In short, contrary to the presumptive attitudes of European settlement, ecosystems in North America were affected profoundly by human activity.

Given this background, religious and philosophically inclined groups have been slow to embrace sustainability as a meaningful way of thinking about nature or the natural environment. While some have worked to meld sustainability with theological traditions, very little of this thought has percolated down to the level of lay churchgoers or environmental activists. In many respects, religion remains as it was when White wrote his provocative article. Many religiously inclined people are totally unengaged in environmental consciousness and are comfortable with the exploitation of natural resources. A smaller contingent stresses the sanctity of nature and promotes conservation or preservation of natural systems. The emphasis on stewardship in Western faith traditions does suggest how humanity and natural systems can (and, if one is religious, should) be interconnected in a manner that is harmonious and mutually beneficial. This tradition of religious scholarship is moving toward incorporating sustainability.

What are religious groups doing to promote sustainability?

Religious organizations may be especially receptive to the idea that sustainability means more than addressing environmental problems. The traditional religious emphasis on charity and assistance to the poor or needy finds expression in those aspects of sustainable development that were targeted toward people in less industrialized parts of the world. Although religious missionaries have a decidedly mixed record in their impact on both the environment and the peoples they have tried to help, the idea of bringing aid to the less fortunate has deep roots in almost all religious traditions. As such, the notion that people currently living in poverty should be blocked from opportunities to develop the natural resources under their control is inconsistent with religious morality. Religions have also supported some form of duty to unborn generations, so the notion of development that allows current generations to meet their needs "without compromising the ability of future generations to meet their own needs" resonates with many religious communities. In short, for religious groups, the primary meaning of sustainability aligns with longstanding duties to help the poor.

A few exceptional trends are worth noting. The Parliament of World Religions is a loosely organized assembly of representatives from different faith traditions. When the first meeting convened in 1893, the focus was on relieving religious conflict, especially when it was a source of violent confrontations. The role of religion and faith in addressing environmental issues gradually assumed a larger role in their deliberations, and the 2018 meeting in Toronto included sessions dedicated to the exploration of an interfaith conception of sustainability. Here, the traditional religious ethic of helping the poor joins with the ecumenical commitment to world peace in promoting a comprehensive vision of sustainability that is compatible across the full range of religious perspectives.

At a more abstract level, the theologian John B. Cobb worked closely with the economist Herman Daly to develop an Index of Sustainable Economic Welfare (similar to the GPI described in chapter 5). Together they published *For the Common Good*, providing an early and influential statement of ideas central to the emergence of sustainability as a broad social movement.

Individual churches, synagogues, and mosques developed and promoted activities to discourage waste, support more efficient use of resources, and promote ecological integrity. More concretely, Interfaith Power and Light is a US-based group that describes itself as making a religious response to global warming. They coordinate action at the local level through regional chapters and maintain a website with numerous suggestions for action in both policy advocacy and individual behavior. Promoting change toward alternative food systems has been an especially significant activity for many religious organizations. Hazon, a Jewish organization that promotes sustainability in the United States, supports farmers' markets and engages in community-supported agriculture as ways to promote more sustainable practices. In summation, religiously affiliated groups are engaged across the spectrum of activities intended to promote sustainability.

What are the arts, and how do they relate to sustainability?

The classical arts are sculpture, painting, poetry, drama, and music. Today, scholars who study the arts include a variety of creative activities: photography, film, literature, digital media, and performance. Increasingly, cooking and eating have important artistic dimensions. Creative design brings art to the construction of many everyday objects, from toasters to clothing or automobiles. Here, practical needs and manufacturing techniques combine with aesthetic principles. Products that exhibit style and beauty enhance the quality of life and allow users to create a personal environment attuned to their own tastes. While many artworks are made and consumed at the scale of an individual household, others become part of the landscape, forming the infrastructure for social interaction. Architecture combines creative design with principles of engineering to develop new types of buildings and urban infrastructure.

Defined broadly, sustainable art is creative activity that takes its relationship to social and ecological domains into consideration throughout each phase of the creative process. Like businesses, artists may choose their materials with an eye to the resources that go into their work or that will be consumed in later phases of appreciation by an audience. In addition, some artists have chosen subject matter that reflects a particular interpretation of sustainability. They may also hope to communicate this interpretation to their audience or aim to invoke a reflective reaction from those who see, hear, or otherwise experience the work. Such works often focus on a specific dimension of sustainability such as pollution, climate change, or social justice. When a work of art aspires to some sense of permanence—an oft-sought trait in the classical arts—taking sustainability into account requires the artist to consider the demands that will be placed on future generations. This has led some artists to see works that are explicitly designed for a short life span, perhaps to dissolve into the environment, as more sustainable. Here, sustainability is reflected in the very idea of artistic creation.

What is sustainable architecture?

Architecture is a field within the arts where sustainability has had a widespread impact. Architects became aware that built infrastructure has dramatic and long-lasting impacts on broader ecosystems. All building materials use natural resources, and designers practice conservation by avoiding woods that grow slowly, come from areas already under pressure from deforestation, or come from threatened species such as rosewood or mahogany. In such cases, the artistic challenge is to reimagine the aesthetic possibilities for using materials that can be derived from a regenerative ecosystem. William McDonough is a Virginia architect who popularized this approach to sustainable architecture with labels such as "cradle to cradle" and "the circular economy," both stressing design principles that reuse and recycle materials.

By the 1980s, architectural theorists were already reacting against the influence of regimented design principles associated with architectural modernism. The architecture of the twentieth century emphasized function and the use of commodity building materials such as concrete, steel, and glass. The International Style was an aesthetic behind the monolithic appearance of commercial buildings constructed from the 1920s well into the 1970s. One reaction to such design principles was regionalism, a theory of architecture that attempts to reflect historical and ecological elements of the place where a structure is erected in its design. Regionalists stress both the use of local materials and designs that help to sustain a sense of place. Pliny Fisk III is a Texas architect who combines technical aspects of sustainable design with the aesthetic principles of place-oriented design.

How do the visual arts practice sustainability?

Paintings, sculpture, and other forms of artistic creation intended to be appreciated solely for their aesthetic qualities have also embraced the parallel approach to sustainability exemplified by architecture. One track discourages materials (such as paints) that contribute to the toxic waste stream and promotes the use of recycled materials. Here, the contribution is to minimize the material burden that artistic activity places upon already stressed ecosystem services. The sustainability of the artwork may not be apparent to a viewer. A second track incorporates waste materials into artwork in a highly visible manner, demonstrating how materials that would be discarded can be repurposed through creative activity. This type of creative reuse of materials in the waste stream is sometimes known as "upcycling."

As with architecture, the incorporation of waste materials into artworks follows trends that predate the emergence of sustainability as a theme for artistic practice. As early as 1917, Marcel Duchamp exhibited "ready-mades," which were ordinary objects that Duchamp appropriated and displayed as works of art. Later, pop art also celebrated ordinary objects, though pop artists did not reuse these objects. The sculptor César Baldaccini (1921–1998) created a body of work almost entirely from discarded materials. Dozens (if not hundreds) of twenty-first-century artists are following his lead, drawing inspiration from the process of creating an aesthetic object from waste materials. Although one might question whether this can have a significant impact on the amount of materials going into the waste stream, the art has the dual purpose of calling the viewer's attention to the fact that artwork has been made from the tons of plastic, metal, and organic waste materials that are disposed of on a daily basis. The art thus conveys a message of sustainability at the same time that sustainability or reuse is incorporated into the artists' practice.

Do other arts also practice sustainability?

Yes. The two-pronged answer given for architecture and visual arts can be traced through virtually any area of the arts. Creative artists and the publishers or production companies that support them are looking for ways to reduce the footprint of artistic production. Producers of printed texts are shifting toward less toxic inks and recycled paper, as well as to formats (such as e-books) that eliminate the need for printing altogether. Musicians have moved strongly toward digital media and away from the vinyl, tape, and compact disc formats that require energy and resources for their production and generate waste. Musicians are also discussing formats that would reduce the environmental impact of performances. Indeed, the Internet has thoroughly transformed the material reproduction of many artistic products, allowing widespread access at very low cost in terms of natural resource consumption or environmental impact.

Second, writers, musicians, and other artists have drawn upon the work of their forebears to craft new works that engage sustainability themes in a contemporary manner. Let's focus on music first. Nature themes became common in nineteenth-century European music, both as the subject matter of songs and operas and through symphonic pieces that mimicked the sound of a natural environment. Song lyrics took a decidedly environmentalist turn in the 1960s and 1970s for rock and roll bands and folk artists. In 1971, the Beach Boys released "Don't Go Near the Water," a new twist on their odes to surfing. That same year, John Prine released "Paradise," decrying the environmental damages from strip mining in the coalfields of Kentucky. Instrumental music took a turn toward sustainability with Philip Glass's compositions for Godfrey Reggio's 1982 film *Koyaanisqatsi*, a montage of insults to the environment keyed to the meaning of the Hopi word that served as the film's title: life out of balance. Activism through music has long addressed broader social and economic themes and promoted what some would see today as a form of social sustainability. Ludwig van Beethoven's Ninth Symphony includes a choral section stressing the future brotherhood of all peoples. African American artists developed coded critiques of slavery and Jim Crow through their field hollers and blues. Pete Seeger's "Where Have All the Flowers Gone?" is a well-known antiwar anthem. In "The Ghost of Tom Joad," Bruce Springsteen revisited an iconic emblem of class struggles, calling attention to the plight of the poor and disenfranchised. Hip-hop artists continue the tradition of popular music that integrates social and environmental themes.

However, it would be contrary to the overall message of this book to suggest that there are no costs to these transformations in artistic production. Successful musicians still travel constantly, and the carbon footprint from their performance work is significant. It also takes materials to create the original products, and it takes energy to run the servers that power the Internet. While it is reasonable to speculate that the digital revolution has made the arts more sustainable, readers should bear three points firmly in mind. First, everyone should temper enthusiasm by noting that the impacts of artistic production pale in comparison to that of the industrial economy writ large. Second, there is virtually no data or life-cycle analysis that supports claims one way or the other about the contributions to sustainability being made by such transformations. People should not simply assume that no hidden impacts challenge the sustainability of current practices. Consistent with hip-hop music's focus on racial injustice and poverty, everyone should remember that sustainability is not just about the environment. Do the arts perform vital social functions that strengthen our institutions and enrich our lives? Many people think that they do. Anyone should consider how these social systems are affected by technical changes intended to reduce art's environmental impact before they conclude that these changes are progressive.

What other arts reflect sustainability in their subject matter?

Many works of literature, theater, television, and film reflect themes related to sustainability as well. Indeed, there must be thousands of novels, essays, poems, plays, and films that articulate the importance of preserving nature and maintaining the integrity of natural ecosystems. As with the arts and architecture, contemporary works that communicate the importance of sustaining the environment have many precursors. Romantic poets such as William Wordsworth (1770–1850) and Percy Bysshe Shelly (1792–1822) stressed nature's potential for stirring an emotional, aesthetic reaction. Henry David Thoreau's *Walden* appeared in 1854. The twentieth century saw a continuation of this nature writing in the works John Muir (1838–1914) and Aldo Leopold (1887–1948). Contemporary writers in this tradition include Barbara Kingsolver and Bill McKibben, both of whom pay explicit attention to the systems orientation of sustainability in their nature writings.

Film has been an especially effective medium for artists hoping to craft works that center on the unsustainability of current practice and the need for reforms. Hundreds of films shine light on environmental issues. The best known include

An Inconvenient Truth, Al Gore's 2006 documentary on climate change, and *Avatar*, James Cameron's 2009 science fiction exploration of environmental exploitation and the quest for sustainability. *Honeyland*, a documentary about a Macedonian wild beekeeper and the challenges of balancing ecological and social dynamics, received Oscar nominations in 2020. In short, an ample supply of artistic works has engaged the topic of sustainability, so we cut the answer to this question short and encourage you to simply search online for your favorite art form (drama, poetry, etc.) with the word sustainability. You won't come up empty-handed.

However, one weakness of the Internet search strategy is that the results will probably focus extensively on environmental concerns and less so on the intractable social concerns. It will miss a theme that we've noted several times already: previous generations have been concerned about the sustainability of their way of life and have identified threats to it in the form of conflict, economic collapse, and social injustice. Industrialization and human population growth are two of many factors that have made environmental threats more critical for sustainability than they might have been for artists like Wordsworth or Beethoven. In addition, relying on an artistically effective message can be misleading in the long run. *Too Hot to Handle*, another climate change documentary released in the same year as *An Inconvenient Truth*, was more faithful to climate science than the more well-known film. *Avatar* has been criticized for using a white savior narrative that undervalues the knowledge and capacity of marginalized and repressed peoples. Nonetheless, these more popular films did raise awareness. An artistic performance thus may be more like an informative indicator than a reliable one (see chapter 4). A full appreciation of the role that the arts can play for sustainability requires that you see how the environmental dimension interacts with socioeconomic subsystems that have long been associated with the continuous reproduction of cultural resources and progressive responses to the distress and suffering of impoverished or marginalized parties.

Can religion and the arts help us achieve sustainability?

We confess right up front that we do not have the answer to this question. It is important to point out that many ecologists are now turning to the religious views of traditional peoples to gain insight into the way that human societies have regulated their impact on larger ecosystems. In addition, more and more scientists and policymakers are coming to the view that society will never make progress on the wicked problems of sustainability without the arts. Art—including popular forms such as movies and television—stimulates an emotional reaction while it communicates ideas. Religion engages us with our deepest concerns. Both may be able to generate a way of knowing that is more immediate than the sciences, and that helps people relate the topic of sustainability to their own experience. And let's face it: much of the technical detail in scientific models intended to deepen our understanding of stocks, flows, and feedbacks in complex systems (like the climate) is just plain boring. Yet people quite rightly expect that communicators will engage them respectfully, rather than adopting a position of superiority. It is at least arguable that religion and the arts offer forums in which everyone is on an equal footing in the discussion. So our answer to this question is: let us hope so.

9

SUSTAINABILITY

WHAT EVERYONE NEEDS TO ASK

Can I do anything to improve sustainability?

Our answer is yes, but far be it from us to tell anyone what they need to do. First, recall chapter 1. People inquire about the sustainability of many practices and processes. Individuals involved in planting a garden or running a business probably had ideas for sustaining it before reading this book, and specialized expertise will help more with things like that than anything we have written. We talked about operating a business to help people get some idea of how sustainability can be understood across multiple levels of a hierarchy, but we didn't offer any practical advice to business operators. We'll interpret this question to be asking about large comprehensive systems. This would include ecological processes that operate at a planetary scale, as well as the way that these systems are integrated with systems of human practice that encompass all manner of socioeconomic and cultural activity. Our answer is still yes, you can do things that strengthen the sustainability of these big systems; hierarchy means that subsystems can affect more comprehensive systems. However, the complexity of these systemic interactions means that the first step is to approach sustainability with humility. It is easy to be wrong, and it can be costly too.

All the previous chapters in this book include at least some reference to values conflicts that lie beneath important questions about sustainability. The way people live and the choices they make each day do advance (or threaten) sustainability, but they are personal decisions. It is only in combination with the impact of choices that everyone else is making that personal decisions affect sustainability. It would not be incorrect for an individual to say that the impact of his or her conduct is insignificant, if their actions are considered in isolation from what everyone else is doing. That said, none of us should think that our decisions don't have an impact of one sort or another. Our point is that we aren't going to offer suggestions for how people should rearrange their lives or make huge changes or look for magic bullets. But what we can offer is a framework through which readers can think critically about key questions they should be asking themselves and others about how they can contribute to improvements—whether they are focused on business, or the environment, or social justice, or governance.

Put simply: learn to think like a mountain. That was conservationist Aldo Leopold's metaphor for considering how a hierarchy of stocks, flows, and feedbacks among predator and prey species creates a system that can either be stable, resilient, and adaptive or tend toward collapse. We have shown how systems thinking can help us understand sustainability in other ways. In addition to thinking about ecosystems, you can use systems thinking to make better sense of sustainability in a business or economic development context, as well as in the evaluation and maintenance of governance institutions or human activities such as science, art, and religion. All of these domains operate within and depend upon more comprehensive social practices, and social practices occur within a larger system of environmental stocks, flows, and feedbacks. Being able to recognize such interdependencies is a crucial step in taking action to advance sustainability.

We have qualified the importance of systems thinking in two important ways. First, people need to understand how they may be undermining the strength and resilience of subsystems that must be sustained to support everything else (that is, systemic interactions at higher levels of the hierarchy). The impacts can even accelerate when hierarchies are taken into consideration. The dilemma of climate change, for example, is that a system of flows and feedbacks is contributing to a gradual increase in the average planetary temperature. Growth in this key stock triggers feedback that affects melting of the polar ice caps. Depletion of *that* stock reduces the earth's ability to reflect solar radiation back out into space. That, in turn, increases average temperature, which starts the cycle again. Although this process of mutually reinforcing feedbacks was set off by human activity, it has created a new set of systemic interactions that, if left to run, will continue to accelerate, causing impacts on sea level, agricultural production, and the frequency of catastrophic weather events such as droughts, floods, hurricanes, and (counterintuitively) bouts of extreme cold.

Second, readers should recognize that there can be stable and resilient institutions that reinforce social patterns of interaction that no one wants to sustain. Everyone can and should ask whether things like persistent poverty, social injustice, and structural racism might be all too sustainable. If you imagine the amount of poverty, injustice, or racially distributed inequity in our society as a stock, you can then ask: What are the flows that might increase or decrease this amount? Are there feedbacks from other quantities that affect these inflows and outflows? Finally, you can ask whether feedbacks seem to maintain these unwanted stocks at levels that, in turn, stimulate *other* systemic interactions (such as increases in resentment, hostility, or anger) that affect the sustainability of governance institutions (see chapter 7) or the social capital that is crucial for a business's ability to operate (see chapter 2). Identifying all of these connections is admittedly not easy, and there are likely to be disagreements. Nevertheless, beginning to think of these social phenomena in systemic terms is one of the big things that everyone can do to improve sustainability.

Another way to approach sustainability with humility is to take the goal of promoting sustainability seriously, but don't beat yourself (or anyone else) up too much when further considerations introduce further complications. There are so many things people can do that it is difficult to know where to start. We have shown that sustainability is thought about in many different ways. The most typical proposals for action emphasize the environmental dimensions of sustainability. But our hope is that previous chapters suggested other kinds of actions as well. For example, business owners benefit by understanding the connections between the structure of their businesses and the structure of the larger economy. Everyone should consider how systemic oppression overrides the best intentions of individuals. Our goal for this chapter is to help frame the way someone might begin to explore what actions or changes in behavior could have a positive impact on the kinds of systems and social goals they care about. Exploring means asking questions and deliberating on the answers. Addressing the wicked problems embedded in discussions of sustainability means that there is no single, easy answer to any particular question. The risk is that indecision stymies us in the face of such complexities. Hence our suggestion to start out by thinking like a mountain. Fine, but then what?

How about starting small for practice?

There is wisdom in starting with the sustainability of your household budget. Just balancing your budget is working toward the financial sustainability of your personal life. The sustainability of your personal way of life can be evaluated in terms of whether your expenditures are, on average, less than your income. Unless you are so wealthy that you do not need to think about what you spend, you need to balance your expenditures so that, on average, they do not exceed your income. Failing to do that leads to bad things like loss of credit, being forced to dedicate a significant chunk of income to paying off debts, harassment by creditors, personal bank-ruptcy, and, in extreme cases, the loss of a home or automobile and even imprisonment. This thought provides an opening to broader kinds of systems thinking when individuals think be-yond their own household to the system of social and environ-mental interactions at a larger scale.

This is also the way we began our explanation of a sustain-able business back in chapter 2. We went on to illustrate how a firm's ability to make a profit depends partly on what is hap-pening in the larger economy. The firm is a system embedded in a larger economic system composed of other firms, as well as individuals who are buying or selling goods and services. Think of your household in exactly the same way. Many of us have little direct or immediate control over our income. Our income is from our jobs, and it will be fairly steady from week to week or month to month. People maintain a sustainable household by ensuring that expenses stay within that amount. However, that means householders depend upon employers to make sure *their* budgets are sustainable. People who depend on a pension or support from a family member must hope that those sources are solvent. In either case, the household budget exists in a hierarchy that will eventually extend all the way up to the general economy.

This point is so important that we will repeat it several times, trying to say the same thing in different ways. Being able to recognize how stocks (the budget) and flows (income and ex-penditures) exist within a hierarchy of other stocks and flows is the first step toward systems thinking. What everyone needs to know about sustainability is that wherever or however we are defining it, the continuance of any process or practice de-pends on the larger systems that encompass it. In the context of business, as the economy weakens, businesses struggle to maintain profits. In your household, the larger balance of in-come and outgo affects the sustainability of sub-budgets, like paying your child's allowance or, if they are off at college, their living expenses. Such relationships go both ways: the sustainability of a comprehensive system depends upon the sustainability of subsystems. When the US housing market bubble burst in 2005-06, the millions of mortgage holders who were unable or unwilling to continue making their mortgage payments contributed to waves of repercussions through the banking industry and the larger economy and, ultimately, the 2008-09 recession. The point here is that making any system or practice more sustainable depends upon seeing how it is connected to other processes and how they affect its basic stocks and flows. Thinking about your household budget as part of a hierarchy of systems is good practice for thinking about other systems you affect and are affected by.

Are there simple guidelines for evaluating the sustainability of day-to-day activities?

We already noted that any household can think about their financial sustainability simply in terms of income and expenses. In chapter 6, we noted how sustainability came to be defined in terms of sustainable development. Here the question was "How can global society maintain economic growth without depleting the natural resource base, leading to a decline in the quality of life for future generations?" Almost every time you spend money, you are contributing to GDP, which is the leading indicator of economic growth, so in one sense simply spending money is contributing to the development half of what sustainable development requires. However, our understanding of *sustainable* development tells us that making consumption choices that do not help deplete natural resources or, barring that, that affect them less than some alternative choice helps to promote sustainability.

There are a million little ways you can promote sustainability in this context, but they can be lumped together into three main categories: (1) making choices that use fewer resources in a direct sense; (2) choosing goods that are more efficient in their use of resources than alternatives; and (3) making choices that do not threaten fragile systems and that, whenever possible, promote the resilience of these systems. Sometimes these rules can be easy to follow, but other times the more sustainable path is extremely difficult to ascertain. And then there are times when doing something you thought would contribute to sustainability turns out to make things worse. Also, in important if less obvious ways, our actions may promote, or fail to promote, sustainability that has little to do with natural resources or fragile ecosystems. Before considering some of the hard cases, it is helpful to say a bit more about the three main categories.

What can people do to consume less?

There are some no-brainers in this category. Start with a focus on water. Water is plentiful in a few regions, but on a global basis, access to clean, potable water or water for food production is highly constrained. What can you do to consume less? Take shorter showers. It saves water as well as the energy needed to heat it. Don't leave the water running while you brush your teeth. Don't leave the hose running in your garden or driveway while you pause from work to take a phone call. Although water can be cycled through the earth's ecosystems and used again and again, only so much of it is usable at any given time or place. In many places, water consumption is tied inextricably to energy use and to the health of aquatic ecosystems, so water conservation can have broader sustainability benefits.

Energy may be the most crucial resource that can be conserved by mindful consumption. Drive less. You save on finite fossil fuel supplies, and you reduce your contribution to climate-forcing emissions. Turn down your thermostat to reduce the amount of energy you use heating your home, and turn it way down when you are not there. Energy conservation is a twofer. First, most of the energy people use for transportation or for heating comes from sources that are nonrenewable, so conserving them promotes sustainability. And second, burning those energy sources contributes to global warming. So using less energy is one of the main things that everyone can do.

The general advice here is to avoid wasteful consumption. In chapter 4, we introduced the term throughput. Reducing consumption will also reduce the amount of stuff you send into the waste stream. This means reducing the use of resources up front and reducing the risk of further problems from waste disposal facilities or when waste streams are ill-managed. Choosing products that are more durable and don't need to be replaced as often is one example. Don't buy stuff that you are only going to throw away. Think carefully about packaging and whether you can choose products that have less. When you run to the store for that loaf of bread, do you really need the cashier to put it in *another* bag? Most people have opportunities to make significant contributions to sustainability simply by planning their day in a way that will reduce the amount of energy and material goods they consume, and doing so definitely counts toward sustainability.

How does using more efficient products contribute to sustainability?

The second rule of thumb is to choose goods that make more efficient use of resources. Note that the efficiency we are talking about is measured in terms of a product's impact on the natural resource base. If you want a cup of coffee, for example, you may have a choice between options that consume more or less energy and other resources (soils, water, paper, etc.) to grow and process the coffee itself, to heat the water, and to deliver it to you. You may drink coffee from a one-time-use cup, for example, or from a mug that can be reused an indefinite number of times (though, of course, this use *will* consume water when the mug is washed). The less impact on resources per cup of coffee, the more efficient a given option is. The idea is that you promote sustainability when you choose products that make more efficient use of the resources that go into making them.

It can be difficult to know which of your choices is more efficient, but there are quite a few products on the market that have efficiency ratings. Automobiles now come with ratings that estimate gas mileage, as well as more comprehensive ratings (in some cases) that estimate total environmental impact, including during the manufacturing process. Even when such information is not on the sticker, it may be available online. The US Environmental Protection Agency has developed a Green Vehicle Guide that helps shoppers estimate the relative efficiency of different options. More broadly, household appliances such as refrigerators or washing machines usually come with labels that estimate their energy efficiency, and similar labels or information sources on energy consumption can be found for television sets and other electric appliances.

Even more generally, manufacturers now label many products as "green," "eco-friendly," or "sustainable." This often means that the maker has adopted a more efficient process for the product's manufacture and/or distribution, though you can't tell from such vague claims what their contribution to sustainability actually is. There is nevertheless an argument for selecting products that are claimed to be more efficient because you are sending a signal to producers that you place a value on sustainability and may even be willing to pay more for it. This is not to say that you should be foolish about spending your money this way, but even symbolic gestures can contribute to sustainability by increasing a company's incentive to compete with other firms on the sustainability of their product lines. You also need to watch for greenwashing, but more on that a bit later.

How can consumer purchases promote resilience?

This is where things start to get tricky. Nevertheless, we can identify a few clear-cut examples. There are international conventions governing trade in items made from endangered plant and animal species. Consumers who want to support sustainability will be scrupulous in observing these rules. The Marine Stewardship Council (MSC) has developed standards to curtail the collapse of aquatic species due to overfishing. Just like purchasing an energy-efficient refrigerator, looking for seafood that bears the MSC logo will promote sustainability.

Many organizations develop guidelines or standards for all the various dimensions of sustainability. Learning about them may be getting a little deeper than most day-to-day consumers want to go, but it cannot hurt to know that they are out there. The International Organization for Standardization (ISO) is one of the largest and most comprehensive (discussed in chapter 7). ISO has standards that relate not only to resource consumption but to other aspects of sustainability such as treatment of workers, fair terms of trade, and non-exploitation of poverty-stricken labor or vendors. Finding an ISO code for a particular aspect of its sustainability may require getting down into the fine print on a product, but motivated consumers can do it.

Can my consumption choices really make a difference to sustainability?

This question takes up a general challenge to consumer-oriented contributions to sustainability, and our answer to this question is yes. There are at least three ways in which your decisions about what products to buy make a difference. First, consumers make direct contributions to key stocks and flows. The previous sections describe some of these.

Second, consumer purchases influence the decisions that businesses make by affecting demand for their products. As noted in chapter 2, some businesses are now incorporating sustainability into their decision-making, but consumers can help encourage such decisions. How they can do this is complex and heavily debated. This may get a little deeper into the weeds than some readers want to go, so if you just trust us on that point, skip to the last paragraph under this question. Businesses make decisions based on their understanding of the demand for products of a given sort. Demand is partially derived from actual consumer behavior, but it also includes an extrapolation to what consumers would do if the supply and price of products were different. Many economists would advise us against the thought that a single consumer's choice has any direct impact on producer decision-making, but they would also say that every consumer's actual purchase behavior is reflected in the estimate of demand that does determine a producer's choices.

Here is how we think you should make sense of this. It is wrong to think that your consumption purchases cause a direct change in producer behavior, but it is correct to think that they are part of the information that producers are using to make their own choices about what and how to produce. This means that if you deliberately choose products that promote sustainability and avoid products that are detrimental to sustainability, you are effectively telling producers that evaluating sustainability as a component of demand is consistent with their desire to make a profit. This may or may not cause them to favor more sustainable products or manufacturing processes; that is but one variable among a whole host of factors they must consider. Nevertheless, your conduct has become one component in the information that producers use to assess demand. If producers ascertain that they can recover the costs they incur in making a product more sustainable, they are likely to do it. If they determine that they can gain an advantage over their competitors by doing so, they are almost sure to do it. The purchases that you make become part of the information they are using to make these decisions, especially when you show that you are willing to pay extra for products that are more sustainable.

Finally, humans are social animals, and the conduct of one person influences the behavior of others. Human nature being what it is, people tend to behave somewhat alike, especially among groups that associate together frequently or view each other as neighbors and friends. This generalization should be taken as just that—a generalization; we can all cite examples where it does not hold. Nevertheless, if a shopper starts to notice more and more people picking up a particular product in the grocery store, they may at least try it too. If they think it is the normal thing to do, it may rise to the level of a default option. There may not be much thought behind this process. If habits and patterns are accompanied by occasional conversations in which people explain how they believe that their purchases promote sustainability (perhaps through one of the ways discussed above), peer pressure and reflective thinking will combine to influence more and more people to go along. In comparison to the previous two ways your behavior helps promote sustainability, this one sounds downright mysterious, but the influence of simple conformism gets greater and greater with each person added. Malcolm Gladwell pressed this point in his book *The Tipping Point*. It should not be neglected as a reason to think that your individual behavior can (eventually) make a difference.

Should I be concerned about greenwashing?

Yes, but don't use this as an excuse for doing nothing. Going all the way back to chapter 2, you'll recall that greenwashing occurs when a company advertises a product as more friendly toward the environment without really doing much to ensure that it actually is. This is a deceptive practice, and that alone is a reason to condemn greenwashing. Another reason to watch for this kind of deception is because you may strive to spend your hard-earned dollars to promote sustainability but find you haven't helped at all. And worse, with a bit of research, your purchasing power might have made a difference if you had used it elsewhere. Try to watch for opportunities to expose examples of greenwashing and follow up on them. However, don't become so cynical that you just revert back to buying the cheapest, sweetest-smelling, or most attractive version of the product, like you might have done before you learned about sustainability. As we explained back in chapter 2, even when you get suckered by greenwashing, your purchase still tells the makers of that product that you care about sustainability. That gives businesses an incentive to do better and to figure out ways to boost your confidence that they are really doing what they claim. The message for businesses is that pursuing sustainability matters; there are countless examples of companies that are capitalizing on this fact.

Should I recycle? And if so, what?

Reducing and recycling are two frequently cited recommendations for promoting sustainability. The questions we've considered so far emphasize reducing the impact of consumption by consuming less or by consuming products that make more efficient use of resources. Recycling suggests that consumption can be made more resilient when people convert the waste stream (an outflow from consumption) into an inflow for some other production process. As a general idea, this goes right to the heart of resilience, but as with many other things, it can be devilishly difficult to realize in practice. Not so long ago, the only way to recycle many household items in our community was to load them into your own car and haul them down to the recycling center. This meant that many people were driving extra miles in their personal vehicles (not to mention using time that could be spent on things that make larger contributions to sustainability), and those extra miles were releasing climate-forcing gases into the atmosphere. Now many people have curbside recycling, which is easier, but there are still trucks driving around picking up this household waste. Either way, recycling reduces the amount of waste material going into landfills or incinerators, both of which introduce their own difficulties into the equation. The answer to the recycling question therefore depends on how the life-cycle calculations pencil out.

By this point, no one will be surprised to learn that the life-cycle calculations are complex. In short, the life cycle for recycling of glass items appears favorable presently, while paper items have potential that is currently more difficult to realize. It depends in part on the energy needed to convert waste into a usable product. Also, some processes of converting waste to a usable product require a great deal of water. In addition, the direct cost of converting a waste product may exceed the cost of making the product from a nonrecycled resource. Thus, if there are no policy incentives for adjusting these costs, a company may be reluctant to put themselves at a competitive disadvantage relative to other firms by using a costlier stream of inflows in their manufacturing process. And, of course, the recycled materials need an end use, a buyer. In some cases, international trade relations even affect recycling markets. There is more work to do on realizing recycling's potential contribution to resilience. However, we think the decision to recycle sends messages similar to those we discussed for consumption decisions. By showing your support for recycling, you are demonstrating that sustainability matters to you, even if you aren't quite sure where your materials are ending up. You are sending a message to companies and to political leaders that sustainability is a worthwhile social goal. You are doing *something*.

How do I think about other choices?

By describing how you can use your power as a consumer or your recycling choices to promote sustainability, we don't mean to suggest that choices are easy or straightforward. In fact, in some cases, going down this path is downright hard. Consider a common sustainability-related choice: paper or plastic?

This is a question people face regularly (unless public policies in your locale have decided the question for you). Maybe the answer is neither: use a reusable cloth bag. But let's back up a step. Principles of recycling and life-cycle analysis lie at the root of the question. Where does the bag figure in the larger system of environmental stocks and flows? Most paper or plastic bags eventually end up in the trash after shoppers bring them home, so let's start the analysis with that flow. Paper bags, it is argued, are easily decomposed. Plastic, on the other hand, decomposes very slowly, so plastic bags hang around in landfills for ages. Not so fast, though—research has shown that very little decomposition actually happens in land-fills. Have you seen those pictures of the fifty-year-old hot dog dug out of a landfill? And in 1992, the *New York Times* pub-lished a story about an archaeologist who found a serving of guacamole in a landfill right next to a newspaper from 1967. If your waste goes into an incinerator, you might want to explore how it handles emissions. Burning plastic emits more harmful chemicals than burning paper. If your waste is loaded onto a barge and hauled out into the ocean for dumping, then decom-position does matter. Paper will decompose; plastic will hang out and threaten aquatic life in the water and other species if it finds its way to shore. Of course, both paper and plastic bags are recyclable, if that option is available to you.

What about flows into the stocks of bags? Paper bags are made from wood, so trees have to be cut down to make them. This decreases the stock of trees, and since they turn carbon dioxide into oxygen and the wood itself captures carbon and removes it from the atmosphere, everyone should want to maintain this stock. On the other hand, trees are a renewable resource. Most of the trees cut down for paper production come from forests that are managed for pulpwood (the wood that is used to make paper) with reforestation an important part of the process. Plastic bags are made from a byproduct of petroleum refining, which represents a flow from a nonre-newable resource: fossil fuels. As a byproduct, plastic bags are not directly implicated in using up a nonrenewable resource. What about the production processes themselves? How much pollution is generated by paper making? In the making of plastic bags? How much soil erosion or habitat disturbance is caused by harvesting trees? By drilling for and pumping oil?

Maybe this paper or plastic conundrum is at the heart of the push for reusable bags. However, if your reusable bags are cotton, have you looked into where and how the cotton was grown and processed? Much of cotton production involves intensive irrigation and pesticide applications. Some reusable bags are plastic, but they don't create the inflow to the waste stock that single-use plastic bags do. Using reusable bags is an example of reducing throughput—using fewer raw materials and producing less waste. But putting fresh meat into a reusable bag is a safety hazard, because microbes will stay there and grow long after you have put your purchase in the refrigerator. You need to take the stocks and flows of microbes into account too.

Paper or plastic (or reusable) is a good example of a sustainability question. We hear lots of other similar questions: Should I give up flying? Should I stop using plastic straws? Should I become a vegan? What our little lesson indicates is that the answer depends on perspective and values, as well as systems analysis. If you think that the plastic bags are likely to wind up as hazards for fish or wildlife, you may think that it is better to use paper, despite the effect that it has on the demand for wood products. Even if you can and do bring reusable bags to the grocery store, you would be well advised to consider putting meats and other products that support microbial growth into a disposable container. You could, of course, wash your cloth bag at a temperature high enough to kill the microbes, and the energy needed to heat the water and considerations of where the water goes when it leaves your house become part of the life-cycle analysis. This is the point in the discussion where some people are inclined to throw up their hands and give up. We hope you will not do that, but we do think that these complexities illustrate why everyone should give others some leeway when they are trying to do things that are more sustainable. One person's capacity will vary from another's, and there are reasons to be tolerant of disagreement, even as everyone works to improve sustainability. Intolerance is probably not (socially) sustainable.

Why isn't everyone concerned about sustainability?

This is a great question. We've already talked about how much personal values come into play in the kinds of choices and decisions we are considering. No one can possibly attend to all the ways in which values may differ in this context. However, everyone should be aware of some important behavioral drivers. First, no one should be surprised if individuals or families who are struggling to make ends meet do not spend much time or money on helping to protect ecological or social systems. Political oppression exacerbates this. One writer described the problem as a finite pail of woe: people can only worry about so many things at once. Remember Maslow's hierarchy of needs. This theory tells us that people work to satisfy basic physiological needs first—like food, shelter, clothing, and so on. Then they focus on needs related to personal health and safety. Next in the list is needs related to a sense of belonging or connection to others, followed by self-esteem, status, and recognition. Self-actualization needs are at the top. Worrying about environmental protection, for example, is a luxury that individuals cannot necessarily afford until they are at a higher level in the hierarchy. Some researchers have argued that concern for the welfare of humanity—sustainability, say—comes as part of self-actualization.

Second, evidence is growing that the wealthiest members of society are least likely to be negatively affected by damages to ecological or social systems. Thus, it is argued, sustainability concerns don't weigh heavily on those who have access to the resources needed to insulate themselves from potential harm. Recent news stories have told how Brazil's economic recession has caused an escalation in the number of homeless people among the poor in Sao Paulo; in contrast, wealthy residents are merely reducing the rate at which they use helicopters to travel around the city. A recent study in the United States showed that flooding associated with severe storm events—which are expected to increase in number and severity with climate change—disproportionately affect the urban poor. Homes, schools, and medical centers in low-income neighborhoods are more likely to be damaged and less likely to be prioritized for repairs and reconstruction. Meanwhile, the poor are less likely to have needed insurance and more likely to be unemployed when affected businesses are closed.

Together, these two phenomena seem contradictory. Those on the lower end of the income spectrum cannot worry about sustainability, and those on the upper end don't need to. These are generalizations, of course. Nevertheless, the first reminds us to be careful about judging others' decisions; the second reminds us to be sensitive to the responsibilities that should come with good fortune. Sustainability does demand a little ethics, after all.

How can I help beyond being a more responsible consumer?

There are probably a million ways to help, but we will stress two. One is proactive: it requires action on your part that will make changes in current practice. The other is more passive: don't do things that undermine the work that other people are doing, even if it is not consistent with your approach.

In chapter 1, we suggested that individuals can work through the political process to encourage decisions by businesses, policymakers, and other leaders that promote sustainability. In chapter 7, we argued that sustainable governance, or the ability of governments to accomplish the things citizens want them to do, depends in part upon our willingness to help with the steering and rowing of governance processes. So looking for opportunities to be politically active or to be part of collaborative efforts to achieve important environmental and social goals is something everyone can do. There are many ways to do this. If you are active in a political party, you can work to align its platform and policy goals more closely with sustainable practice, and you can encourage others in the party to evaluate party goals from a systems perspective. If you are involved in any kind of citizens' action group, you can suggest efforts to align the group's objectives and strategy with all dimensions of sustainability. Well-known environmental objectives, like reducing climate-forcing emissions or cleaning up pollution, might be the specific focus of these efforts, but other systems-oriented objectives, like strengthening the resilience of governance mechanisms or supporting scientific research, are equally legitimate. Sweeping environmental laws enacted during the 1970s and associated regulations imposed in response to public concern and outrage resulted in vast improvements in environmental quality. But people face new problems regularly, and backsliding is a constant risk. Dealing with disturbingly resilient problems like poverty, racism, and inequality of opportunity requires large, concerted efforts. In 1964, President Lyndon Johnson declared a war on poverty in the United States. Many of those initial programs were ill-conceived, but in the intervening years, nutrition assistance, cash assistance, and early childhood education programs, among others, have improved the sustainability of industrial society. Individuals' charitable gifts and volunteer activities help, but this remains a difficult battle.

Some of the most significant opportunities for political action are at the local and regional levels, however. Modifying local infrastructure to decrease energy use (especially for transportation purposes) and thinking more systemically about the way that zoning and commercial development cause feedback to other social quantities beyond economic growth are two examples. The sustainability of local governance processes is especially sensitive to sound fiscal policies, so promoting sustainability may involve careful, systems-based analysis of revenue streams for city, county, and state programming. Volunteer activities (such as a food bank or charitable fundraising) can also have feedbacks for other local efforts. Thinking through all these activities from a systems perspective promotes sustainability.

In addition, no one should overlook opportunities to have an influence in the organizations in which they participate either. Corporations and small businesses alike are looking for more sustainable practices, so depending on your position in the firm, you may be able to suggest or implement a more sustainable path. Visionaries in some industries are leading the charge to reduce greenhouse gas emissions—both because their business relies on threatened ecosystem services and because they strive to be good corporate citizens. Similarly, organizations from churches to garden clubs can undertake activities that raise awareness of environmental injustice or that aid others in a transition toward greater efficiency in resource use. Businesses, nonprofit organizations, and faith-based initiatives are leading important charges to address inequities in access to nutrition, education, housing, and safety. The point of all this is simply to emphasize something that might seem obvious to some people but nonetheless tends to get lost in the shuffle of endless advice on consumption decisions. People are not going to achieve sustainability through better shopping alone; more conventional types of cooperative activity like political organization, workplace reform, and general social action are also needed.

The passive contribution you can make (i.e., things you should *not* do) can be summed up in a sentence: don't be a finger-wagging moralist when others are trying to do something that helps or avoid things that hurt. Given the complex ways that systems interact, everyone should expect that people who occupy different social positions may fail to see some things that they see clearly, but people should also remind themselves that others may see things that they have overlooked. Things that may seem simple or easy to one person may be virtually impossible for someone in a different position to undertake. You should therefore be tolerant of others who are doing less than you are or who are making their contribution to sustainability in ways that you do not understand or agree with. The first order of business is to get systems thinking and sustainability on the front burner, and everyone should expect that people will approach this in different ways.

This is not to imply that you should never talk about sustainability with friends and family. Notice that we've just pointed out how exchanging ideas about the systems in which our activities are embedded is one of the ways that people can help. However, the detailed questions we have mentioned already (Should I give up flying? Should I stop using plastic straws? Should I become a vegan?) rapidly became points of divisiveness, largely because people took up the change in their own behaviors as a moral duty they believed everyone else was obliged to follow. Here it is important to remember that very few aspects of sustainability even require unanimous agreement in the first place. They may need a significant number of people to change, but they don't need everyone. What is more, moralizing has never been the most effective way to convince anyone of your point.

Is sustainability just a passing fad?

This is the disheartening question that cynics and disheartened people are likely to ask. Pursuing sustainability certainly feels like a fad sometimes. In this book, we have covered how sustainability means different things depending on whether one is looking at a business, at ecosystems, at social justice, or at processes of governance or research. Sustainability is not a fad in any of these domains. Yet we have seen an endless parade of magic-bullet actions recommended in the name of sustainability. Curbside recycling, switching from plastic to paper, becoming vegan, drinking Fair Trade coffee, giving up your bottled water, biking to work—the list is long, and there are reasons to think that all of them help. People encourage each other to do them for a while, but eventually the enthusiasm wears out. While there is usually a rationale for them, they often conceal complications that make them seem less compelling on closer inspection. Is there nothing more to sustainability than a series of fads that go viral and then die out? This endless cycle can make it seem like there is nothing really there. But this is not the real problem. The real problem lies in thinking all of the items on your to-do list are magic bullets or that the goals they promote are morally mandatory.

The solution to this problem is to recall the complexity created by interacting systems. While an action can increase sustainability by making more efficient use of some resource, that action can have rebound effects that do just the opposite. Sometimes the rebound is in systems (like the global climate system) that most of us do not understand in the first place. Seeking sustainability requires you to remain faithful to the objective, even while you remain open to the possibility that any particular strategies might prove to be less effective than you originally thought, and sometimes they are just wrong altogether. More than twenty years ago, the authors of an article titled "Zen and the Art of Climate Maintenance" wrote, "Sustainability is about being nimble, not being right." Put another way, we all, every one of us, still have a lot to learn.

Everyone *should* listen to experts. The science of sustainability can produce insight into the structure of systems, and it can help identify leverage points where interventions that get us off an unsustainable path are most effective. At the same time, everyone needs to understand that experts are learning too. That means that sometimes they discover better options or learn that an early hypothesis is actually wrong. However, you will benefit more from listening to experts when you do not simply take what they say at face value but instead try to fit it into an understanding of the economy, ecosystems, and our social life based on system dynamics.

Does the complexity of sustainability mean we should just throw up our hands in despair?

This is the reverse way of asking the very first question in this chapter, "Can I do anything to improve sustainability?" The answer there was yes, and the answer here is no. However, the challenges can be so daunting that anyone is likely to feel overwhelmed and depressed occasionally. We had this feeling when we learned that washing fleece jackets made from recycled plastic bottles (in fact, washing any synthetic fabric) introduces microplastic pollutants into the environment. On days like that, it seems we just can't win. One lesson to take from those days is that there is always more to learn. People should avoid overconfidence in the quick fix. The other lesson is that anyone can profit from mistakes, sometimes through unexpected ways.

Organic agriculture illustrates both points. Farmers formed organic farming organizations in the 1970s to share information and to develop standards that would help them market their products. These early projects sought both environmental and economic sustainability. Organic producers were pursuing environmental goals by limiting their use of pesticides and promoting soil fertility. They were also hoping to establish a niche market where they could get a higher price for their produce than they would in the commodity markets dominated by large industrial farms. As the organic farming movement developed, definitions for organic production were merged, eventually resulting in highly consistent rules for the US Department of Agriculture Organic Standard and standards for "biologic" production in Europe and elsewhere.

Organic foods have proven to be a hit with an economically significant minority of consumers. Producing organic products from fruit and vegetables to dairy, meat, and even fiber goods from cotton has proven to be highly profitable. Were the original sustainability goals achieved? Organic production reduces the use of many chemical inputs, but some have argued that any such contributions to environmental sustainability are offset by the fact that organic farming requires more acres to produce the same amount of food. Perhaps conserving land for biodiversity would be a more worthwhile environmental goal.

However, if the environmental contributions of organic agriculture are contested, the socioeconomic case seems clear. Organic farming is so profitable that very large, industrially organized farms now dominate the sector, just as they dominated commodity markets before the organic farming movement started. Small farmers have found that selling in farmers' markets or direct to restaurants (who may in fact demand organic products) is a more reliable way to achieve economic sustainability. In addition, note that many consumers purchase organic products because they believe that they are safer, but food safety was never an explicit goal of the organic agriculture movement. The sustainability goals sought by the organic standard have been twisted through the process of building a market for them. Studies continue to show that the health benefits from eating organic food are marginal at best. Sustainability in food systems thus turned out to be more complicated than anyone suspected.

At the same time, the commercial success of organic production has had good effects on the food system when taken as a whole. The fact that a small movement of farmers could promote a sustainability goal and a significant number of consumers would respond to it was noticed by industrial farmers, agricultural policymakers, and the major firms that supply farmers. The rise of the organic movement joined with other feedbacks to increase the amount of attention given to agriculture's impact on the global environment. The economic success functioned as a message to food industry firms (not just farmers) that consumers might be willing to pay more for products that were more sustainable. Thus, even if one questions whether the movement for organic farming achieved its sustainability goals, it nonetheless has helped promote sustainability.

The bottom line, then, for what you can do to promote sustainability is to follow Aldo Leopold's advice. Learn to think like a mountain. Remember that you are affected by and can affect all kinds of systems, and you will learn about them as you go. Asking questions about sustainability is asking questions about how things work and what you can do to promote the continued functioning of important social and ecological systems. We should all do that.

NOTES

Chapter 2

1. Fishman, C. (2006). *The Wal-Mart effect: How the world's most powerful company really works—and how it's transforming the American economy.* Penguin Books.

2. Will, G. (2015, April 15). "Sustainability" gone mad on college campuses. *Washington Post.* Retrieved from https://www.washingtonpost.com/opinions/sustainability-gone-mad/2015/04/15/f4331bd2-e2da-11e4-905f-cc896d379a32_story.html?utm_term=.6fddf8f2f76e.

3. How GM saved itself from Flint water crisis. (2016, January 31). *Automotive News.* Retrieved from https://www.autonews.com/article/20160131/OEM01/302019964/how-gm-saved-itself-from-flint-water-crisis.

Chapter 4

1. At the AirNow website (https://www.airnow.gov), maintained by the EPA, you can look at current air quality for cities around the United States as well as cities around the world where the United States has a consulate. Ozone and fine particulate matter are the two constituents reported, as these are the most linked to human health.

2. You can use tools on the Carbon Footprint website to calculate your carbon footprint https://www.carbonfootprint.com/.

3. This information is available on the Nature Conservancy website at https://www.nature.org/en-us/about-us/who-we-are/how-we-work/working-with-companies/companies-investing-in-nature1/coca-cola-company/.

Chapter 5

1. Technically, GDP accounts for *final* goods and services. That means, for example, that wheat sold by a farmer to a mill and flour sold by a mill to a bakery aren't included directly in GDP calculations. They are considered intermediate products included in the bread sold by the bakery to the consumer. This convention avoids counting things multiple times.

Chapter 6

1. Sugrue, T. J. (2014). *The origins of the urban crisis: Race and inequality in postwar Detroit.* Princeton University Press.

Chapter 7

1. See the Bertelsmann Stiftung's Sustainable Governance Indicators website, https://www.sgi-network.org.
2. The Fragile States Index is available online at https://fragilestatesindex.org.

FURTHER READING

Grober, U. (2012). *Sustainability: A cultural history* (R. Cunningham, Trans.). Green Books. (Original work published 2010)

Jacques, P. (2015). *Sustainability: The basics*. Routledge.

Meadows, D.H. (2008). *Thinking in systems: A primer* (D. Wright, Ed.). Chelsea Green Publishing.

Walker, B., & Salt, D. (2006). *Resilience thinking: Sustaining ecosystems and people in a changing world*. Island Press.

Walker, B., & Salt, D. (2012). *Resilience practice: Building capacity to absorb disturbance and maintain function*. Island Press.

INDEX

For the benefit of digital users, indexed terms that span two pages (e.g., 52–53) may, on occasion, appear on only one of those pages.